Lectures on Modular Forms

Joseph Lehner

DOVER PUBLICATIONS, INC.
Mineola, New York

Bibliographical Note

This Dover edition, first published in 2017, is an unabridged republication of the work originally published by the National Bureau of Standards, Washington, D.C., in 1969. The material originally appeared as Volume #61 in the National Bureau of Standards' Applied Mathematics series.

International Standard Book Number:

ISBN-13: 978-0-486-81242-7
ISBN-10: 0-486-81242-1

Manufactured in the United States by LSC Communications
81242101 2017
www.doverpublications.com

Foreword

The theory of modular forms, which received its systematic treatment at the hands of Felix Klein in the late 19th century, has had a wide-ranging impact on mathematics and its applications. Groups of integral matrices (with their fundamental applications to crystallography), conformal mapping, ergodic flows, symmetries on manifolds, and number theory are among the many disciplines to which this subject has been applied. In recent years the fundamental work of C. L. Siegel, who generalized the concepts to functions of several complex variables and applied them to the theory of quadratic forms, has aroused new interest in modular forms.

There is a need for an exposition that starts with the classical material and leads gradually to modern developments. These lectures make few demands on the maturity of the reader, and their study will acquaint students and workers in the field with the important basic ideas, and bring them to the point where they can understand and appreciate the current research in this subject.

Lewis M. Branscomb
Director

Contents

Lectures on Modular Forms

This book is an expository account of the theory of modular forms and its application to number theory and analysis. The first chapter defines modular forms and develops their most important properties. The second and third chapters introduce the Hecke modular forms. A Hecke form f is a simultaneous eigenfunction of a family of linear operators T_p, where p runs over the positive primes. The eigenvalue corresponding to T_p is the pth Fourier coefficient $a(p)$ of f. The Fourier coefficients of f are multiplicative, i.e., $a(mm) = a(m)a(n)$ when m and n have no common divisors. The intimate connection of these results to the theory of certain zeta-functions is indicated in a note to chapter 3. The fourth chapter is devoted to the automorphisms of a compact Riemann surface. It is shown what groups of linear-fractional transformations with coefficients in a finite field can be the group of automorphisms of a compact Riemann surface. In the last two chapters congruences and other arithmetic properties are developed for the Fourier coefficients of Klein's absolute modular invariant. Analogies with the Hecke theory as well as with the Ramanujan congruences for the partition function are discussed.

Keywords: Automorphism; finite field; genus; Hecke operator; Klein's invariant; linear transformation; modular form; Riemann surface.

Introduction

The simply-periodic functions such as the sine, cosine, and exponential, are familiar to every reader. Such functions are invariant under a linear transformation $\tau' = \tau + b$, where b is a constant. The functions that are invariant under a linear-*fractional* transformation

$$\tau' = \frac{a\tau + b}{c\tau + d}, \quad ad - bc = 1$$

also play a prominent role in many branches of mathematics. Of these an important subclass is the class of *modular* functions, which we obtain by requiring that a, b, c, d be integers. A *modular* form has roughly the same relation to a modular function that a differential (or a power of a differential) has to a function. These lectures are concerned with modular forms and their applications to number theory and analysis.

The reader is required to know only the elements of complex variable theory, group theory, and number theory. Explanation of difficult points is provided in Notes collected at the end of each chapter; they are referred to in the text by superscripts, for example,[1]. These Notes also contain further elaborations of the theory and its connection with other mathematical theories.

Reference is made to two of the author's books for certain proofs and detailed discussions. These are:

A Short Course in Automorphic Functions, Holt, Rinehart, and Winston, New York, 1966
Discontinuous Groups and Automorphic Functions, American Mathematical Society, Surveys No. 8, Providence, 1964

and are referred to as Short Course and Surveys, respectively. The reader will find that if he is willing to accept certain statements on faith, he will not need to consult these works much.

The material of this volume was presented at the Canadian Mathematical Congress, Algebra Seminar, held at York University (Toronto) in August 1967. The author is indebted to Margaret Ashworth (now Mrs. Millington), who made a careful reading of the original draft.

This book is dedicated to the memory of HANS RADEMACHER.

Joseph Lehner

Institute for Basic Standards
National Bureau of Standards

and

Department of Mathematics
University of Maryland

Chapter I

Modular Forms

1 The Modular Group and Some Subgroups

Throughout these lectures we shall use the letter Γ to denote the modular group, which is the group of linear-fractional transformations

$$\tau' = \frac{a\tau + b}{c\tau + d}; \; a, b, c, d, \in Z, ad - bc = 1. \tag{1}$$

Here Z is the set of rational integers. In group theory this group is referred to as $LF(2, Z)$. Though the transformations (1) are the object of our study, it is much more convenient to use matrices. The group*

$$SL(2, Z) = \left\{ \begin{pmatrix} a & b \\ c & d \end{pmatrix} \middle| \, a, b, c, d, \in Z, ad - bc = 1 \right\}$$

is not quite isomorphic to Γ, but we have

$$\Gamma \simeq SL(2, Z)/\{I, -I\}, \; I = \begin{pmatrix} 1 & 0 \\ 0 & 1 \end{pmatrix}. \tag{2}$$

*For typographical convenience we shall often write $(a\ b \mid c\ d)$ in place of $\begin{pmatrix} a & b \\ c & d \end{pmatrix}$ when it appears in the text.

We regard Γ as the matrix group $SL(2, Z)$ in which each matrix is identified with its negative.

We shall be concerned with certain subgroups of Γ, all of finite index:

$$\Gamma_0(n) = \left\{ \begin{pmatrix} a & b \\ c & d \end{pmatrix} \in \Gamma \,\middle|\, c \equiv 0 \pmod{n} \right\},$$

$$\Gamma^0(n) = \left\{ \begin{pmatrix} a & b \\ c & d \end{pmatrix} \in \Gamma \,\middle|\, b \equiv 0 \pmod{n} \right\},$$

$$\Gamma(n) = \left\{ \begin{pmatrix} a & b \\ c & d \end{pmatrix} \in \Gamma \,\middle|\, \begin{pmatrix} a & b \\ c & d \end{pmatrix} \equiv \pm I \pmod{n} \right\},$$

$$\Gamma' = \text{commutator subgroup of } \Gamma,$$

and some others. Here it is necessary to be careful about the relation of the matrix to the transformation. Clearly, if $-I$ is in the matrix group, (2) holds; otherwise the two are isomorphic. The first case occurs in the first 3 groups listed above, but $-I \notin \Gamma'$. In general, we shall make no distinction between the matrix and transformation groups, trusting to the context to keep things straight. For applications to number theory the group $\Gamma_0(n)$ is especially important.

We shall denote by G an arbitrary subgroup of finite index in Γ. The group G is *discrete*, that is, it contains no infinite sequence of distinct matrices that converges to the identity matrix. Almost all of what we do goes over with minor changes to discrete groups of linear-fractional transformations with real coefficients provided they are finitely generated, but we shall not be concerned with the more general case except in Chapter IV.

The geometric theory of linear-fractional transformations is essential in our work. We shall deal only with linear-fractional transformations with real coefficients and determinant 1; call the group of such transformations Ω. An element of Ω maps the upper half-plane **H** on itself and the real axis on itself, and conversely any transformation with these properties can be written with real coefficients and determinant 1 and so belongs to Ω. We also have the classification into 3 types (disregarding the identity):

$$\begin{aligned} &\text{elliptic if } |a+d| < 2 \\ &\text{parabolic if } |a+d| = 2 \\ &\text{hyperbolic if } |a+d| > 2. \end{aligned} \tag{3}$$

The interpretation as noneuclidean motions is familiar.[1] (cf. ch. I notes, No. 1, p. 16). An elliptic transformation has two complex-conjugate fixed points, one lying in **H**; a hyperbolic transformation, two real (distinct) fixed points; a parabolic transformation, a single real fixed point. In particular, elliptic elements of G can only have traces 0 or ± 1; the former are of order 2, the latter of order 3, and these are the only orders possible

[1] Throughout this work these arabic footnote numbers refer to the corresponding chapter notes given at the end of each chapter.

for elements of G. We shall consider that all groups of linear-fractional transformations act on $\mathbf{H}$ or on the real axis.

The points $a, b, \in \mathbf{H}$ are said to be *G-equivalent* (or simply *equivalent*, when G is understood) if $Va = b$ for some element $V \in G$. By this equivalence relation $\mathbf{H}$ is partitioned into mutually disjoint equivalence classes or *orbits*

$$Gz = \{Vz | V \in G\}.$$

The concept of orbit leads to the *fundamental region*, on the one hand, and to the *Riemann surface*, on the other. These are both realizations of the orbit space $\mathbf{H}/G$, defined as the set of distinct orbits of G. $\mathbf{H}/G$ is the space obtained by identifying points in $\mathbf{H}$ that are G-equivalent.

To realize the orbit space in $\mathbf{H}$, select one point from each orbit and call the union of these points a *fundamental set* for G (relative to $\mathbf{H}$). Since we wish to deal with nice topological sets, we modify this concept slightly and define a *fundamental region* R_G to be an open subset of $\mathbf{H}$ which contains no distinct G-equivalent points and whose closure contains a point equivalent to every point of $\mathbf{H}$. That fundamental regions exist for the groups of interest to us admits a simple proof (cf. Gunning, ch. I* or Short Course, p. 57). Fundamental regions for Γ and $\Gamma(2)$ are shown in the figures; notice that they are actually regions (i.e., connected), which is not required by the definition. There are, of course, many fundamental regions; in particular, $V(R)$ is one if R is, where $V \in G$. The collection of regions $\{V(R) | V \in G\}$ form a network of nonoverlapping regions which, with their boundary points, fill up $\mathbf{H}$. Examples of these striking geometric configurations may be found in many books.

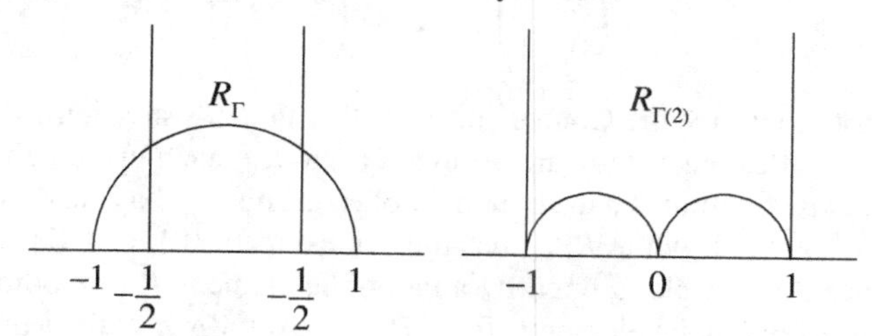

Let A, B be sets in which a multiplication of elements is defined. Write

$$C = AB$$

if the set of products $\{ab | a \in A, b \in B\}$ is the set C. Write

$$C = A \cdot B \tag{4}$$

if for each $c \in C$ we have *uniquely* $c = ab$, $a \in A$, $b \in B$. Thus, with C a group, A a subgroup, we have (4), where B is a system of right representatives

*Gunning, R.C. 1962. Modular Forms. Princeton University Press, Princeton, NJ. 96 pp.

Theorem 1. *Let $G = H \cdot A$, where H is a subgroup of finite index in G. Then*

$$\operatorname{Int} \bigcup_{a \in A} a\bar{R}_G$$

is a fundamental region for H. ($\bar{R}$ is the closure of R.)

The almost evident proof can be found in Short Course, theorem 6D, p. 61f. This fundamental region may not be connected; however, connected fundamental regions do exist for all subgroups.

It is possible to select the fundamental region so that it has other desirable properties. The sides of such a fundamental region are arranged in *conjugate pairs*, the two sides of a pair being equivalent by a group element. These conjugating transformations generate the group.

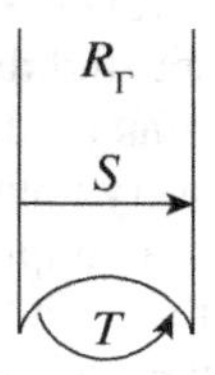

For example, in R_Γ, we regard the arc of the circle as consisting of two sides separated by the point i. Then the vertical sides are mapped into each other by S, the curved sides by T, where

$$S = \begin{pmatrix} 1 & 1 \\ 0 & 1 \end{pmatrix}, \quad T = \begin{pmatrix} 0 & -1 \\ 1 & 0 \end{pmatrix} \tag{5}$$

are standard notations. (Short Course, p. 37.) The *vertices* are defined to be points of $\bar{R}_G$ where two sides meet; they are arranged in *cycles*, each cycle being a complete G-equivalence class of points on the boundary of R_G. Thus R_Γ has the following cycles: $\{\infty\}$, $\{i\}$, $\{e^{2\pi i/3}, e^{2\pi i/6}\}$; while $R_{\Gamma(2)}$ has the cycles: $\{\infty\}$, $\{0\}$, $\{-1, 1\}$ – see figure on p. 3. If one vertex of a cycle is fixed by a parabolic element P, the other vertices are also fixed points of parabolic elements, for VPV^{-1} fixes $V\alpha$ if P fixes α, and VPV^{-1} is also parabolic. Then the cycle is called parabolic. Similarly for elliptic fixed points, and it is clear that every vertex of an elliptic cycle is fixed by an elliptic element of the same order, which is called the order of the cycle. It can happen that the vertices of a cycle are not fixed points at all; then we say the cycle is *accidental* (or *unessential*). Such a cycle can be avoided by a different choice of fundamental region, whereas this is not possible with fixed-point cycles. Obviously an elliptic cycle must lie entirely in $\mathbf{H}$, whereas the vertices of a parabolic cycle are all real (including possibly ∞). The sum of the angles at the vertices of a cycle is 0, 2π, or $2\pi/k$, according as the cycle is parabolic, accidental, or elliptic of order $k > 1$. In R_Γ the cycle $\{\infty\}$ is parabolic, while the remaining cycles are elliptic of order 2 and 3, respectively. In $R_{\Gamma(2)}$ the cycles are all parabolic. (Short Course, p. 39ff.)

If we identify points (necessarily on the boundary) of $\bar{R}_G$ that are G-equivalent we obtain an orientable surface. Thus for R_Γ we get a sphere with one point removed (one "puncture"); for $R_{\Gamma(2)}$ we get a sphere with 3 punctures. The genus of the surface (i.e. the number of handles) is in both cases zero. It can be shown that this always happens: the R_G of every subgroup G of finite index in Γ becomes, on identification of G-equivalent boundary points, a surface of genus $g \geqq 0$ with $t \geqq 0$ punctures (Short Course, ch. III, sec. 1). The integer g can be computed from Euler's formula applied to R_G:

$$c - n + 1 = 2 - 2g, \tag{6}$$

where c is the number of cycles and $2n$ the number of sides. The number of punctures t is simply the number of parabolic cycles in R_G.

Let us calculate the genus of Γ. Here $n = 2$ and there are 3 cycles; this gives $g = 0$. For $\Gamma(2)$ we have $n = 2$ and 3 cycles; again $g = 0$.

These considerations tie in naturally with the Riemann surface[2] of G. The orbit space $R = \mathbf{H}/G$ can be given a topological and, in fact, analytic structure that makes it a Riemann surface. There is a projection map

$$\sigma;\ \tau \to G\tau$$

from $\mathbf{H}$ to R which identifies G-equivalent points:

$$\sigma_0 V = \sigma \text{ for } V \in G, \tag{7}$$

and which is a local homeomorphism except at the fixed points of G lying in $\mathbf{H}$. Thus $\mathbf{H}$ may be regarded as a branched unlimited covering of $R = \mathbf{H}/G$ with the projection σ. Moreover, R_G is a model of R and the genus of R is the same as the genus of R_G as defined above (Short Course, ch. III, sec. 1). Finally σ is an analytic mapping from $\mathbf{H}$ to R except at the fixed points of G.

Poincaré introduced the area metric[3]

$$\frac{du\,dv}{v^2},\ \tau = u + iv \tag{8}$$

which is invariant under all linear-fractional transformations with real coefficients and determinant $+1$. If we restrict ourselves to fundamental regions that are Lebesgue measurable, it is easy to show that they all have the same area, which is therefore a group invariant and is called the area of the group (Short Course, p. 50). The area under this metric of a circular arc triangle lying in the closure of $\mathbf{H}$ is known to be π minus the sum of its angles (Gauss-Bonnet formula). By triangulation R_G we then find, in view of (6), that

$$|R_G| = 2\pi\left(2g - 2 + \sum_{i=1}^{c}\left(1 - \frac{1}{l_i}\right)\right), \tag{9}$$

where $|R_G|$ is the area of R_G and l_i is the order of the transformation fixing any point of the ith cycle ($=1$ if the cycle is accidental, ∞ if the cycle is parabolic). Thus $|R_G|$ is finite. C. L. Siegel has determined the minimum of the right member of (9):

$$|R_G| \geqq \pi/21. \tag{10}$$

This remarkable but completely elementary result was apparently unknown until the appearance of Siegel's paper in 1945 (Some remarks on discontinous groups,

Ann. of Math.). For groups with parabolic, elements – all the groups G are in this class – we have a stronger result:

$$|R_G| \geqq \pi/3. \tag{11}$$

The minimum is attained by Γ. (Short Course, ch. I, sec. 5.)

From theorem 1 we deduce that if $[\Gamma : G] = \mu$, then

$$|R_G| = \mu|R_\Gamma|, \tag{12}$$

since $|V(R_\Gamma)| = |R_\Gamma|$.

2 Modular Forms

A *modular form* is an analytic function that has a certain type of invariant behavior on Γ or a subgroup of Γ. We recall that the symbol G is used to denote an arbitrary subgroup of finite index in Γ. The simplest behavior would be strict invariance:

$$f(L\tau) = f(\tau) \text{ for all } L \in G. \tag{13}$$

However, this is too restrictive for our purposes. If we consider *differentials* rather than functions, we observe that a consequence of (13) is

$$f'(L\tau)d(L\tau) = f'(\tau)d\tau.$$

Differentials of higher "weight" would be obtained by taking powers; thus, with $g = f'^h$ we get

$$g(L\tau)(dL\tau)^h = g(\tau)(d\tau)^h.$$

Now

$$L'(\tau) = (c\tau + d)^{-2}, \ L = \begin{pmatrix} a & b \\ c & d \end{pmatrix}$$

if we remember that $ad - bc = 1$; the above equation is then

$$(c\tau + d)^{-2h} g(L\tau) = g(\tau), \ L \in G, h \in Z. \tag{14}$$

The well-known Weierstrassian invariants g_2, g_3 from the theory of elliptic functions satisfy (14) with $2h = 4$ and 6, respectively.

To express (14) it is very convenient to introduce the *stroke operator*:

$$f(\tau)|_{-k}L = (c\tau + d)^{-k} f(L\tau), \ L = \begin{pmatrix} a & b \\ c & d \end{pmatrix}. \tag{15}$$

Here L has real coefficients and determinant $\neq 0$; in our applications $k = 2h$ is an even integer and is usually omitted from the notation. Equation (14) is then simply

$$g|L = g, \ L \in G. \tag{16}$$

Observe that

$$g|L_1L_2 = (g|L_1)|L_2. \tag{17}$$

We can also define, for constants α_1, α_2:

$$g|(\alpha_1L_1 + \alpha_2L_2) = \alpha_1 g|L_1 + \alpha_2 g|L_2. \tag{18}$$

Moreover, we shall want g to satisfy some analytic requirements. It would be natural to insist that g be meromorphic, but in our applications only holomorphic functions arise. The second requirement concerns the behavior of g at real rational points (including ∞): we wish g to have a definite behavior as τ tends to such a point. We now make the following definition.

Let $\{G, -k\}$ be the set of functions $f(\tau)$ such that (19)

(i) f is holomorphic in $\mathbf{H}$.

(ii) for each matrix $A \in GL^+(2, Z)$, i.e., with integral entries and positive determinant, $f|_{-k}A$ tends to a definite limit (finite or infinite) as $\tau = u + iv \to i^\infty$ uniformly in every region E_α: $|u| < 1/\alpha$, $v > \alpha > 0$; this limit is independent of u.

(iii) $f|_{-k}L = f$ for all $L \in G$.

If r is a rational point, we define the *value of f at r* as follows: Let V be a matrix of $GL^+(2, Z)$ such that $V\infty = r$; define

$$(f)_r = (f(\tau)|V)_{\tau=\infty} = \lim_{\tau \to i^\infty} (f(\tau)|V). \tag{20}$$

The existence of this limit follows from (19)(ii); that it is independent of V (provided $V\infty = r$) is easily shown.[4] Thus imposition of (ii) guarantees that f has a value at all rational points. Note that

$$(f+g)_r = (f)_r + (g)_r \text{ and } (f|L)_r = (f)_{Lr}$$

The value $(f)_r$ may be infinite. Note that $(f)_r$ is not necessarily equal to $f(r) = \lim_{\tau\to r} f(\tau)$. For suppose $f \in \{\Gamma, -k\}$ and suppose $f(\infty) = 1$. Let r be finite and let $V\infty = r$, $V = (a\,b|c\,d)$. Then $c \neq 0$. So $f(V\tau) = (c\tau + d)^k f(\tau)$; letting $\tau \to \infty$ we have $f(r) = \infty$. But $(f)_r = (f|V)_\infty = (f)_\infty = 1$. However $(f)_\infty = f(\infty)$, since we may take $V = I$.

We call f a modular form of degree $-k$ (actually, a regular modular form). If $k = 0$, we speak of a modular *function*. Restrictions like (ii) are absolutely essential if the theory of modular forms is to have an algebraic character.

Clearly we have

$$f_1 \in \{G, -k_1\},\, f_2 \in \{G, -k_2\} \implies f_1 f_2 \in \{G, -k_1 - k_2\}.$$

In this way we can generate forms of all negative even integral degrees $\leqq -4$ if we have a form of degree -4 and one of degree -6, say.

We shall now define subclasses of $\{G, -k\}$. Let

$$\langle G, -k\rangle = \{f \in \{G, -k\} \text{ such that } (f)_\infty \neq \infty,\, (f)_r \neq \infty \text{ for all rational } r\}. \quad (21)$$

This is the class of *everywhere regular* forms. Next let

$$\langle G, -k\rangle_0 = \{f \in \{G, -k\} \text{ such that } (f)_\infty = 0,\, (f)_r = 0 \text{ for all rational } r\}. \quad (22)$$

These are called *cusp forms*. Clearly $\{G, -k\}$, $\langle G, -k\rangle$, and $\langle G, -k\rangle_0$ are all vector spaces.

It is as well now to give some examples. Let us first consider the modular group Γ. In general we denote by F_α the subgroup of F consisting of all elements of F that fix the point α (the "stabilizer" of α). Considering Γ as a group of linear fractional transformations, Γ_∞ is the infinite cyclic group generated by S. If Γ is regarded as a matrix group, however, then $\Gamma_\infty = \{S, -I\}$. Let us take the latter point of view and consider the following series without regard to convergence:*

$$G_k(\tau, \nu) = \sum_{M \in B} e^{2\pi i \nu\tau}\Big|_{-k} M, \quad k = \text{even integer} \quad (23)$$

where $\Gamma = \Gamma_\infty \cdot B$ and $\nu \in Z$; it is easy to check that the series does not depend on the choice of B. In fact, if $\Gamma = \Gamma_\infty \cdot B'$, then each $b_i \in B$ corresponds to a $b_i' = V_i b_e \in B'$ with $V_i \in \Gamma_\infty$. Hence $V_i = \pm S^{n_i}$ and $b_i'\tau = b\tau + m_i$, so that the term of the series corresponding to b_i is unchanged on going to b_i'. (Actually B runs through a full set of matrices in Γ with different lower row.) It follows that for $V \in \Gamma$,

$$G_k | V = \left(\sum_{M \in B} e^{2\pi i \nu\tau}\Big| M\right)\Big| V = \sum_{M \in B} e^{2\pi i \nu\tau}\Big| MV = G_k,$$

since clearly $\Gamma = \Gamma_\infty \cdot BV$. The series (24) converges absolutely for $k \geqq 4$ (so far as even integral values of k are concerned); for the proof see Short Course, p. 68ff. Absolute convergence permits rearrangement of its terms, which was done implicity above when we replaced MV by M. The convergence is uniform in regions $|u| \leqq \alpha$, $v \geq \beta > 0$, and its terms are holomorphic (k is an integer!), all of which implies G_k is holomorphic in $\mathbf{H}$. What is the situation as $\tau \to i^\infty$? For the matrices M with

*When written out this series would read:

$$G_k(\tau, \nu) = \sum_{(c,d)=1} e^{2\pi i \nu(a\tau+b)/(c\tau+d)}(c\tau + d)^{-k},$$

where $M = (ab|cd)$ is any matrix in Γ having (c, d) as its lower row.

$c \neq 0$, $|c\tau + d|^{-k} \to 0$ while the exponential factor is bounded. In fact, if $\tau = u + iy$, $\tau' = M\tau = u' + iy'$, we have

$$y' = y|c\tau + d|^{-2},$$

$$|\exp(2\pi i\nu M\tau)| = \exp(-2\pi\nu y') \leqq \exp\frac{2\pi|\nu|y}{(cu+d)^2 + c^2y^2}$$
$$\leqq \exp(2\pi|\nu|/c^2y) \leqq \exp(2\pi|\nu|/y),$$

which is bounded as $y \to +\infty$. There are two terms in the series corresponding to matrices with $c = 0$; the matrices may be taken to be I, $-I$ and the terms add up to $2e^{2\pi i\nu\tau}$. Hence $|G_k(\tau, \nu)| \to \infty$ as $\tau \to i^\infty$ when $\nu < 0$, G_k is bounded when $\nu = 0$, and $G_k \to 0$ when $\nu > 0$. Suppose we can establish (ii) of the definition (19). Then in all cases $G_k \in \{\Gamma, -k\}$, and $G_k(\tau, \nu) \in \langle\Gamma, -k\rangle$ if $\nu \geqq 0$, $G_k(\tau, \nu) \in \langle\Gamma, -k\rangle_0$ if $\nu > 0$. However, in the last case G_k may be identically zero! This is in fact one of the outstanding unsolved problems of the subject. It cannot happen, however, when $\upsilon \leqq 0$. For when $\upsilon < 0$, $G_k \to \infty$ as $\tau \to i^\infty$, as we have just seen. When $\nu = 0$, we get

$$(G_k(\tau, 0))_\infty = \lim_{\tau\to\infty} G_k(\tau, 0) = 2 + \lim_{\tau\to i^\infty} \sum_{\substack{(c,d)=1 \\ c\neq 0}} (c\tau + d)^{-k} = 2.$$

The function $G_k(\tau, \nu)$ is called a Poincaré series of degree $-k$ and parameter ν. When $\nu = 0$ we speak of an Eisenstein series. The classical invariants of Weierstrass, g_2, g_3 are, up to constant factors, identical with $G_4(\tau, 0)$, $G_6(\tau, 0)$, respectively.

We must now establish property (ii) of the definition (19). Let

$$A = \begin{pmatrix} \alpha & \beta \\ \gamma & \delta \end{pmatrix}, \quad \alpha\delta - \beta\gamma = n > 0$$

be in $GL^+(2, Z)$. We have

$$G_k|A = \sum_{M\in B} e^{2\pi i\nu A'\tau}(c'\tau + d')^{-k},$$

where $A' = MA = (a'b'|c'd')$, $c' = c\alpha + d\gamma$. When $c' \neq 0$, $\exp 2\pi i\nu A'\tau$ is bounded as $\tau \to i^\infty$ by an estimate similar to the one above; then all such terms tend to 0 as $\tau \to i^\infty$. There can be at most two terms in which $c' = 0$, and then $d' \neq 0$ since $\det A' = n > 0$. The sum of these terms has the value $2d'^{-k}\exp 2\pi i\nu(a'\tau + b')/d'$, which has a definite limit for $\tau \to i^\infty$.

We have therefore proved the existence of nonconstant modular forms of all even integral degrees $\leqq -4$.

Next we consider modular forms on G, which is, as usual, a subgroup of finite index in Γ. The group G contains a smallest translation S^N, for $[\Gamma : G] < \infty$, and $G_\infty = \{S^N, -I\}$. As before we set up the series

$$G_k(\tau, \nu; G) \equiv G_k(\tau, \nu) = \sum_{M\in B} e^{2\pi i\nu\tau}\Big|_{-k} M,$$

where $G = G_\infty \cdot B$. Here B is seen to be a complete system of matrices in G with different lower row. Absolute and uniform convergence still hold for $k > 2$, G_k is holomorphic in $\mathbf{H}$, and

$$G_k | V = G_k \text{ for all } V \in G.$$

The proof of property (19) (ii) is the same as in the case of the full group Γ. We have thus established the existence of nonconstant forms in the classes $\{G, -k\}$ and $\langle G, -k \rangle$ for $k \geq 4$.

The value of G_k at a rational point, i.e., $(G_k)_r$, is given by (20); its uniqueness is established by[4].

3 Modular Forms (continued)

Every modular form has a Fourier series. In fact, the subgroup G has a smallest translation S^N; thus $f \in \{G, -k\} \implies f(\tau + N) = f(\tau)$, which implies

$$f(\tau) = \sum_{n=\mu}^{\infty} a(n) e^{2\pi i n \tau / N}, \tag{24}$$

where $\mu \neq -\infty$ by the finiteness condition (ii) in the definition (19). As a matter of notation we reserve the letter x for

$$x = e^{2\pi i \tau}$$

and then can rewrite the Fourier series as a power series in $x^{1/N}$:

$$f(\tau) = \sum_{n=\mu}^{\infty} a(n) x^{n/N}. \tag{24a}$$

The $a(n)$ are still called Fourier coefficients however. We say f has order μ at i^∞, and we call $x^{1/N}$ a *local variable* at i^∞; it maps a neighborhood of i^∞ onto an open disc about $x = 0$.

Moreover, f has an analogous expansion at each rational point $r = a/c$. Let $V = (ab|cd) \in T$, where we may assume $c > 0$; then $V S^m V^{-1}$ fixes r. If we choose m as the least positive integer such that $V S^m V^{-1} \in G$, then $V S^m V^{-1}$ will generate the stabilizer of r in G. From $f | V S^m V^{-1} = f$ we deduce $(f|V)|S^m = f|V$, so that $f|V$ has the usual Fourier series in $\exp(2\pi i \tau / m)$:

$$f/V = \sum_{n=\nu}^{\infty} b(n) e^{2\pi i n \tau / m}.$$

By $(19, ii)$, $f|V$ has a limit as $\tau \to i^\infty$; therefore, $\nu \neq -\infty$. This may be regarded as the expansion of f at $\tau = r = V\infty$. It is essentially unique, for if $W\infty = r$, $W \in \Gamma$, then $V^{-1}W$ fixes ∞ and so $W = V S^h$. Then

$$f|W = f|V S^h = \sum_{n=\nu}^{\infty} b(n) e^{2\pi n(\tau + h)/m} = \sum b'(n) e^{2\pi i n \tau / m}.$$

To get an expansion for f itself, operate with V^{-1} to obtain

$$(-c\tau+a)^k f(\tau)=\sum b(n)e^{2\pi i n V^{-1}\tau/m}$$

Since $V^{-1}\tau=-\dfrac{d}{c}-\dfrac{1}{c^2(\tau-r)}, \quad -c\tau+a=-c(\tau-r),$
this yields

$$(\tau-r)^k f(\tau)=\sum_{n=\nu}^{\infty} c(n)\exp\{-2\pi i n/c'(\tau-r)\},\ c'>0 \tag{24b}$$

with $c(\nu)\neq 0$.

Let us now return to Γ. The Fourier coefficients of G_k are known, but unless $\nu=0$ the formulas are too complicated to be of much use. However, in the case of $\nu=0$ (Eisenstein series), the Fourier coefficients are readily obtained by standard complex variable methods. We multiply G_k by 2 to make the constant term 1 and call the resulting function E_k. Then[5]

$$E_k(\tau)=1+\frac{(2\pi)^k(-1)^{k/2}}{(k-1)!\zeta(k)}\sum_{n=1}^{\infty}\sigma_{k-1}(n)x^n,\ k=4,6,\ldots \tag{25}$$

where $\zeta(k)=\sum_{1}^{\infty} m^{-k}$ and $\sigma_r(n)=\sum_{d|n} d^r$. In particular,

$$E_4=1+240x+\ldots,\quad E_6=1-504x+\ldots. \tag{25a}$$

Here we see the emergence of number theory, for $\sigma_{k-1}(n)$ is a multiplicative arithmetic function.

We can obtain a cusp form by combining the E_k; thus

$$E_4^3-E_6^2=12^3x+\ldots$$

is not identically zero and certainly belongs to $\langle\Gamma,-12\rangle_0$. We write

$$\Delta(\tau)=12^{-3}(E_4^3(\tau)-E_6^2(\tau)). \tag{26}$$

The function Δ has a remarkable product expansion[6]

$$\Delta(\tau)=x\prod_{m=1}^{\infty}(1-x^m)^{24} \tag{26a}$$

and so is never 0 in **H**. Its Fourier coefficients are denoted by $\tau(n)$, "Ramanujan's function,"

$$\Delta(\tau)=x\sum_{n=1}^{\infty}\tau(n)x^n, \tag{26b}$$

and $\tau(n)$ has the same multiplicative properties as $\sigma_{11}(n)$, as we shall see later.

Except for 0 there are no cusp forms in Γ of degree > -12 and there are no everywhere regular forms of degree -2. To see this we shall mention an important formula connecting the number of zeros and poles of any form f in $\{\Gamma, -k\}$, $f \not\equiv 0$, namely,

$$q+\mu+\frac{r}{2}+\frac{s}{3}=\frac{k}{12}, \tag{27}$$

where q is the number of zeros of f in R_Γ and μ is the order of f at i^∞. One can prove this formula by integrating f'/f around the boundary of the fundamental region and taking advantage of the transformation formula (19)(iii). The integers r, s are explained as follows. Because of the existence of a substitution T of period 2 in Γ fixing $\tau = i$, it is possible to expand f in powers of $t = ((\tau - i)/(\tau + i))^2$:*

$$(\tau+i)^k f(\tau) = t^{-k/4}\sum_{n=\nu}^{\infty} b(n)t^n;$$

$-k/4+\nu = r/2$ is called the order† of f at $\tau = i$. Similarly, the existence of an element TS of order 3 fixing $\tau = \rho = e^{2\pi i/3}$ results in the (generally fractional) order $-k/6 + \lambda = s/3$ at $\tau = \rho$. The factor 1/12 in the right member of (27) is proportional to the area of R_Γ. (See Short Course, ch. II, sec. 3.)

In (27) we have $q, \mu, r, s \geqq 0$. If we set $k = 2$, there is no solution. In the case of a cusp form we must have $\mu \geqq 1$ and therefore $k \geqq 12$. This proves our assertion.

Each everywhere regular modular form f can be written as a polynomial in E_k and Δ, and the formula

$$f = \sum_{\substack{j\geqq 0\\ k-12j\neq 2}} \alpha_j E_{k-12j}\Delta^j, \quad E_0 = 1$$

is easily proved by noticing from (27) that $\mu \leqq k/12$ and that $E_{k-12j}\Delta^j$ has a zero of order j at i^∞. To obtain a cusp form we must have $\alpha_0 = 0$. From this we can count the linearly independent cusp forms:

$$\dim\langle\Gamma, -k\rangle_0 = \begin{cases} [k/12], k \not\equiv 2 \pmod{12} \\ [k/12]-1, k \equiv 2 \pmod{12}, k > 2. \\ 0, k = 2 \end{cases} \tag{28}$$

Obviously

$$\dim\langle\Gamma, -k\rangle = 1 + \dim\langle\Gamma, -k\rangle_0. \tag{29}$$

For a subgroup G there is a formula similar to (27); the left member counts the algebraic orders of the zeros and poles, with orders at the interior fixed points counted

*t is a "local variable" at $\tau = i$; it maps a neighborhood of i in R_G onto an open disk about $t = 0$. Similarly, $((\tau-\rho)/(\tau-\bar\rho))^3$ is a local variable about $\tau = \rho = e^{2\pi i/3}$.

†Remember k is even, therefore r is an integer.

fractionally as above, while in the right member the factor 1/12 must be multiplied by $\mu = [\Gamma : G]$, because of (12). The result is

$$\sum_{i=1}^{c} \nu_i \left(1 - \frac{1}{l_i}\right) + q = \frac{k\mu}{12}, \tag{30}$$

q being the sum of the orders at interior points of R_G. For example, the case of $\Gamma(2)$ gives

$$\nu_\infty + \nu_{-1} + \nu_0 + q = \frac{k}{2},$$

in an obvious notation. This shows there are no cusp forms for $k < 6$.

The case $k = 0$ is important; here we have a *modular function*. Formula (30) says:

Theorem 2. *$f \in \{G, 0\}$, $f \neq$ constant, implies f has the same number of zeros and poles in $\bar{R}_G$.*

This has the crucial consequence:

Theorem 3. *A bounded modular function f is a constant.**

For, if $\tau_0 \in R_G$, the function $g(\tau) = f(\tau) - f(\tau_0)$ has a zero in R_G but no pole. Hence $g \equiv 0$ or $f(\tau) \equiv f(\tau_0)$.

This is the fundamental theorem of the subject and we give another proof. Let $p_1, p_2, \ldots, p_\mu$ be the parabolic vertices of $R_G(\mu = [\Gamma : G])$ and set $f_i = \lim f(\tau)$ as $\tau \to p_i$ vertically. Since f is bounded in $\mathbf{H}$, f_i is finite. Now construct

$$F(\tau) = \prod_{i=1}^{\mu} (f(\tau) - f_i);$$

then $F \in \{G, 0\}$ and $F(p_i) = 0$. Let $M = \sup |F|$ in $\mathbf{H}$. There is a sequence τ_n such that $|F(\tau_n)| \to M$. Because of the invariance of F under G we may assume $\tau_n \in \bar{R}_G$ and that on a subsequence $\tau_n \to \tau_0 \in \bar{R}_G$. By continuity, $|F(\tau_0)| = M$. Hence τ_0 is a maximum point and by the Maximum Principle it cannot lie in $\mathbf{H}$. Thus τ_0 is forced to be some p_i and in consequence $M = 0$. It follows that $F \equiv 0$, or $f(\tau) \equiv f_j$ for some j.

Q.E.D.

Theorem 2 states that a modular function f has the same number of poles in any fundamental region. This number, called the *valence*, is positive unless f is constant. The valence also equals the number of times f assumes an arbitrary value c, for $f(\tau) - c$ is 0 if and only if $f(\tau) = c$, and the number of zeros of $f - c$ is the same as the number of its poles, i.e., the same as the number of poles of f. When the valence equals 1, we say f is *univalent*.

Theorem 4. *There exist univalent functions on G if and only if G is of genus zero.*

The proof depends on the Riemann surface and will be discussed later.

---

*This may be phrased: An everywhere regular modular function (i.e., regular in $\mathbf{H}$ and at the cusps) is constant.

Theorem 4 implies that Γ admits a univalent function. Consider

$$j(\tau) = E_4^3\Delta^{-1} = x^{-1} + c(0) + c(1)x + \dots, \ x = e^{2\pi ir}. \tag{31}$$

Clearly $j(\tau) \in \{\Gamma, 0\}$. Also by (26a) Δ is never 0 in $\mathbf{H}$, hence j has its single pole, of order 1, at i^∞. Therefore j is univalent. By theorem 2 it must have a single zero in $\bar{R}_\Gamma$, which is known to be at $\tau = \rho = e^{2\pi i/3}$. In fact $E_4(\rho) = 0$, for

$$\dot{E}_4(\rho) = E_4(\rho)|_{-4}TS = (\rho+1)^4 E_4(TS\rho) = (\rho+1)^4 E_4(\rho), \text{ and } (\rho+1)^4 = e^{4\pi i/3} \neq 1.$$

From (26), also,

$$j(\tau) - 12^3 = E_6^2\Delta^{-1}. \tag{32}$$

Using the information in (25a) and (26a)* it is seen that *the Fourier coefficients* $c(n)$ *of* $j(\tau)$ *are rational integers*. Their arithmetic properties will form one of the main objects of study in later chapters.

Returning to subgroups of the modular group, let us consider $\Gamma_0(q)$, $q =$ prime, the subgroup defined by $q|c$ in $(ab|cd)$. Of course every form on Γ is a form on $\Gamma_0(q)$, but we are looking for "proper" forms. Consider

$$f_k(\tau) = E_k(q\tau);$$

for $V = (a, b|qc, d) \in \Gamma_0(q)$ we have

$$qV\tau = \frac{a(q\tau)+bq}{c(q\tau)+d} = V_1(q\tau),\ V_1 = \begin{pmatrix} a & bq \\ c & d \end{pmatrix} \in \Gamma$$

and so

$$\begin{aligned} f_k|V &= (qc\tau+d)^{-k} f_k(V\tau) = (qc\tau+d)^{-k} E_k(qV\tau) \\ &= (qc\tau+d)^{-k} E_k(V_1(q\tau)) = (E_k|V_1)_{q\tau} = E_k(q\tau) = f_k. \end{aligned}$$

Since (19)(i), (ii) are trivial in this case, $E_k(q\tau) \in \{\Gamma_0(q), -k\}$. This is a general principle, valid also when q is not a prime. On the other hand,

$$\begin{aligned} f_k(\tau)|T &= \tau^{-k} f_k(-1/\tau) = \tau^{-k} E_k(-q/\tau) \\ &= q^{-k}(E_k|T)_{\tau/q} = q^{-k}E_k(\tau/q) = q^{-k} f_k(\tau/q^2), \end{aligned}$$

so $E_k(q\tau)$ is not invariant under T and therefore is not a form on Γ. Since $\Gamma_0(q)$ is maximal in Γ when q is a prime, $E_k(q\tau)$ is a proper form on $\Gamma_0(q)$. Similarly, $\Delta(q\tau)$ is a proper cusp form of degree -12 on $\Gamma_0(q)$.

---

*It can also be proved directly from (25) and (26), without using (26a), that Δ has integral coefficients $\tau(n)$ with $\tau(1) = 1$.

Finally, we shall note the connection of forms on G with differentials on the Riemann surface $R = \mathbf{H}/G$. Let $\tau \in \mathbf{H}$ and let $q = \sigma\tau \in R$, where σ is the projection map defined near (7). If φ is an analytic function on R, then

$$f(\tau)\varphi(\sigma\tau) = \varphi(q) \tag{33a}$$

belongs to $\{G, 0\}$, for, by (7)

$$f(V\tau) = \varphi(\sigma V\tau) = \varphi(\sigma\tau) = f(\tau).$$

Conversely, $f \in \{G, 0\}$ implies

$$\varphi(q) = f(\sigma^{-1}q) \tag{33b}$$

is analytic on R; note that $f(\sigma^{-1}q)$ is well-defined, despite the many-valuedness of σ^{-1}, because of (7). In somewhat more complicated vein we can treat forms and differentials. Thus if $\psi(q)(dq)^m$ is a differential (of weight m) on R, then $g(\tau) \in \{G, -2m\}$, where

$$g(\tau)(d\tau)^m = \psi(\sigma\tau)(d\sigma\tau)^m, \tag{34a}$$

and conversely $g \in \{G, -2m\}$ yields

$$\psi(q)(dq)^m = g(\sigma^{-1}q)(d\sigma^{-1}q)^m \tag{34b}$$

as a differential of weight m.* The relation between the orders of forms and differentials at a point can be calculated:

$$\begin{aligned} &n(\tau, g) = \eta(q, \Omega), \text{ if } \tau \neq \text{fixed point} \\ &n(\tau, g) = \eta(q, \Omega) + m\left(1 - \frac{1}{l}\right), \text{ if } \tau \text{ is an elliptic fixed point of order } l \\ &n(\tau, g) = \eta(q, \Omega) + m, \text{ if } \tau = \text{parabolic fixed point.} \end{aligned} \tag{35}$$

Here n is the order of g at τ, η is the order of $\Omega = \psi(q)(dq)^m$ at $q = \sigma\tau$, and m is the weight. The reader is urged to observe (35) with particular care; failure to do so can result in the most absurd contradictions in going between $\mathbf{H}$ and R! In particular,

> a cusp form of degree -2 on $\mathbf{H}$ corresponds to an everywhere regular differential (abelian differential of the first kind) on R, and conversely. (36)

It is a theorem that a compact Riemann surface (whether punctured or not) admits a univalent function if and only if it is of genus zero. When this is carried over to $\mathbf{H}$ we get theorem 4, above.

---

*The "local variables" at points of $\mathbf{H}$, discussed above, correspond to the "disk homeomorphisms" on R.

4 Sums of Squares

Let us now apply these ideas to the problem of representing a positive integer m as a sum of s squares. If such representations are properly counted, their number is given by $r_s(m)$, where

$$\Theta_s(\tau) = (\theta_3(\tau))^s = \left(\sum_{-\infty}^{\infty} e^{\pi i n^2 \tau}\right)^s = 1 + \sum_{m=1}^{\infty} r_s(m) e^{\pi i m \tau}. \tag{37}$$

The function θ_3 is a classical θ-function and we have

$$\theta_3(\tau+2) = \theta_3(\tau), \theta_3(-1/\tau) = (-i\tau)^{1/2}\theta_3(\tau)$$

the first equation being obvious, and the second being proved, for example, by the Poisson summation formula or by contour integration. Hence if $8|s$, $\Theta_s \in \{\Gamma_\theta, -s/2\}$, where Γ_θ is the subgroup of Γ generated by $S^2 = (1\,2|0\,1)$ and T. It is easy to prove the two characterizations:

$$\Gamma_\theta = \{V \in \Gamma | V \equiv I \text{ or } V \equiv T \pmod 2\}$$
$$\Gamma_\theta = \left\{\begin{pmatrix} a & b \\ c & d \end{pmatrix} \in \Gamma | cd, ab = \text{even integers}\right\}. \tag{38}$$

From the first of these it is clear that $[\Gamma : \Gamma_\theta] = 3$, since there are 6 classes of modular matrices modulo 2. We have

$$\Gamma = \Gamma_\theta \cdot \{I, S, ST\},$$

which we use to construct R_{Γ_θ}. (It is necessary to shift some regions by S^{-2} to attain this figure.) Since $S^2(-1) = +1$, the cycles are $\{\infty\}$, $\{-1, +1\}$, $\{-i\}$, of which the first two are parabolic, the last elliptic of order 2 (being fixed by T).

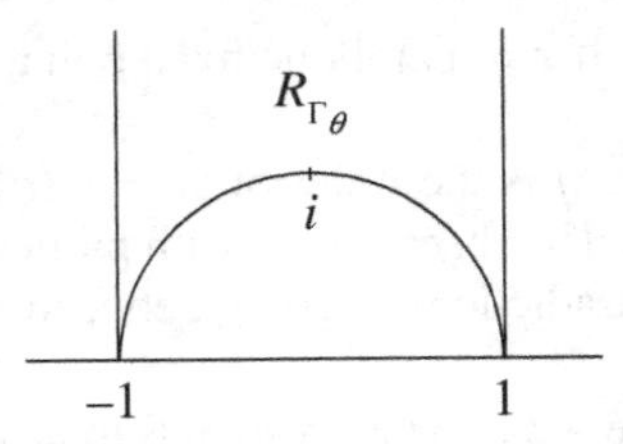

From the definition (37) it is clear that

$$\Theta_s(i^\infty) = 1. \tag{39}$$

By using the known transformations of the theta-functions (see Whittaker-Watson, p. 475) we find

$$\Theta_s | TS^{-1}T = \pm 2 e^{\pi i \frac{s}{4}(\tau+1)} + \dots.$$

Since

$$TS^{-1}T(\infty) = -\begin{pmatrix} 1 & 0 \\ 1 & 1 \end{pmatrix}(\infty) = +1,$$

we get

$$(\Theta_s)_1 = 0. \tag{40}$$

Our next move will be to match up Θ_s with an Eisenstein series

$$G_{s/2}(\tau, 0) = \sum_{M \in B} 1|M,$$

where $\Gamma_\theta = (\Gamma_\theta)_\infty \cdot B$. Let $M = (a\,b|c\,d)$, then

$$M \cdot TS^{-1}T = -\begin{pmatrix} a & b \\ c & d \end{pmatrix}\begin{pmatrix} 1 & 0 \\ 1 & 1 \end{pmatrix} = -\begin{pmatrix} \cdot & \cdot \\ c+d & \cdot \end{pmatrix},$$

and $c + d \neq 0$, since c, d are of opposite parity (cd even, $(c, d) = 1$). It follows that $1|MTS^{-1}T \to 0$ with $\tau \to i^\infty$ and so

$$(G_{s/2}(\tau, 0))_1 \doteq (G_{s/2}(\tau, 0)|TS^{-1}T)_\infty = 0. \tag{41}$$

Now obviously

$$G_{s/2}(i^\infty, 0) = 2; \tag{42}$$

hence (39) to (42) show that

$$\Theta_2(\tau) = \tfrac{1}{2} \cdot G_{s/2}(\tau, 0) + H(\tau), \tag{43}$$

where H is a *cusp* form.

Let us now apply formula (30) to Γ_θ:

$$q + \nu_\infty + \nu_1 + \frac{1}{2}\nu_i = \frac{s}{2} \cdot \frac{3}{12} = \frac{s}{8}.$$

Since a cusp form requires $\nu_\infty > 0$, $\nu_1 > 0$, we cannot have a cusp form unless

$$s \geqq 16.$$

In particular, for $s = 8$ we definitely have

$$H(\tau) \equiv 0$$

and so equating coefficients in the Fourier series of (43):

$$r_8(m) = \tfrac{1}{2}a(m),$$

$a(m)$ being the Fourier coefficients of $G_{s/2}$. It is possible to calculate $a(m)$ since we know the arithmetic characterization of the matrices of Γ_θ. Standard computations give

$$r_8(m) = 16(-1)^m \sum_{d|m} (-1)^d d^3. \tag{44}$$

We can verify that $r_8(m)/16$ is a multiplicative arithmetic function; in fact ($m = 2^\alpha \prod p^e$)

$$r_8(m)/16 = \frac{2^{3(\alpha+1)} - 15}{7} \cdot \prod_{\substack{p^e \| m \\ p>2}} \frac{p^{3(e+1)} - 1}{p^3 - 1}; \tag{45}$$

the first factor is omitted if m is odd. Here $p^e \| m$ means p^e is the exact power of p dividing m.

Chapter I Notes

1. A few remarks on Poincaré's model of hyperbolic geometry are in order. Plane hyperbolic geometry is obtained from euclidean geometry when one replaces the parallel postulate by the following: *Through a given point not on a line there passes more than one line that does not meet the given line.* In Poincaré's model the upper-half plane $\mathbf{H}$ represents the H-plane (= hyperbolic plane), and an H-line is represented by that arc of a circle orthogonal to the real axis that lies in $\mathbf{H}$ (this includes vertical half-lines). We define H-angle measure to be the usual euclidean angle measure. With these definitions it is seen that the axioms of hyperbolic geometry are fulfilled in the model.

   The "rigid motions" of the geometry are those 1–1 mappings of $\mathbf{H}$ into itself that preserve properties of congruence (as stated in the axioms). It can be shown that the rigid motions are the linear-fractional transformations that preserve $\mathbf{H}$, in other words, the elements of $\Omega = LF(2, \text{reals})$. An H-distance can now be defined by means of its differential, which must be Ω-invariant. In fact, let

$$\tau = x + iy, \tau' = V\tau = x' + iy', V = \begin{pmatrix} a & b \\ c & d \end{pmatrix} \in \Omega.$$

   Now

$$d\tau' = (c\tau + d)^{-2} d\tau, y' = y(c\tau + d)^{-2},$$

   as a calculation shows, where we must use $ad - bc = 1$. Thus

$$\frac{|d\tau'|}{y'} = \frac{|d\tau|}{y},$$

   i.e., $|d\tau|/y$ is an invariant differential of length. By integration we obtain the distance formula:

$$d(w_1, w_2) = \frac{1}{2} \log \frac{(w_1 - b)(w_2 - a)}{(w_1 - a)(w_2 - b)},$$

   where $w_1, w_2 \in \mathbf{H}$ and a, b are the real points on the H-line through a, b; a, w_1, w_2, b are in order on the H-line. An H-circle can now be defined as the set of all points at the same H-distance from a fixed point $w_0 \in \mathbf{H}$; it turns out to be a euclidean circle whose center lies vertically above w_0.

Suppose $E \in \Omega$ is an elliptic transformation. It has two fixed points, ξ and $\bar{\xi}$; let $\xi \in \mathbf{H}$. One can prove that each hyperbolic circle about ξ is a fixed circle of E. Putting E in normal form, we get

$$\frac{\tau' - \xi}{\tau' - \bar{\xi}} = e^{i\theta} \frac{\tau - \xi}{\tau - \bar{\xi}}, \text{ where } \tau' = E\tau, 0 \leqq \theta < 2\pi.$$

It is easy to establish that the angle $\tau\xi\tau'$ is equal to θ (map $\tau \to 0$, $\tau' \to \infty$ by a linear transformation, then $\tau' = E\tau$ becomes $w' = e^{i\theta} w$ in the new coordinates). Thus τ has been moved through an angle θ and so we call E a *noneuclidean rotation*.

Next, let H be a hyperbolic transformation of Ω with (real) fixed points ξ_1, ξ_2. If K is any circle through the fixed points (not necessarily orthogonal to the real axis), H moves a point along K by a constant H-distance. We speak of H as a *noneuclidean translation*.

Finally, suppose P is a parabolic element of Ω. If P fixes ∞ (i.e., is a translation), each vertical line is displaced parallel to itself. Similarly, if P has a finite (real) fixed point ξ, an H-line through ξ is shifted by P to another H-line through ξ. Furthermore, if a circle C is drawn in $\mathbf{H}$ tangent at ξ, then P moves a point along C a constant H-distance. Following Coxeter, we refer to P as a *noneuclidean parallel displacement*.

2. The Riemann surface arose originally as the domain space for analytic functions, but independent characterizations were given early in this century by H. Weyl and T. Radó. Let W be a *surface*, that is, W is a connected Hausdorff space and each point of W lies in an open neighborhood D that is homeomorphic to an open subset of the euclidean plane. If D_1, D_2 are overlapping neighborhoods with corresponding homeomorphisms Φ_1, Φ_2, then $\Phi_1\Phi_2^{-1}$ is a mapping (called a neighbor relation) from the complex plane to itself, and we demand that it be directly conformal (analytic with nonzero derivative) wherever it is defined. When these conditions are satisfied, we say W is a Riemann surface. Briefly, a Riemann surface is a surface with directly conformal neighbor relations.

   Now suppose G is a finite-index subgroup of Γ. The points of the orbit space $R = \mathbf{H}/G$ are the orbits $[\tau] \equiv G\tau$, where $\tau \in \mathbf{H}$. The map $\sigma : \mathbf{H} \to R$ defined by $\sigma(\tau) = [\tau]$ identifies G-equivalent points of $\mathbf{H}$, i.e., $\sigma \circ L = \sigma$ for $L \in G$. In Short Course, pp. 117–121, it is proved that R can be given a topology that makes it a Riemann surface. Without going into detail here let us see what the neighborhood structure of R is like.

   As a model of R we can take any set that contains exactly one point from each G-orbit, i.e., a fundamental set. An equivalent procedure is to use a fundamental region R_G together with its boundaries but we identify boundary points that are equivalent under G. Thus, in the case of Γ, we could use the closed standard fundamental region $\bar{R}_\Gamma$ but identify the vertical sides and the circular arcs meeting at the point i. Call such a set R'_G. We can now operate entirely in R'_G.

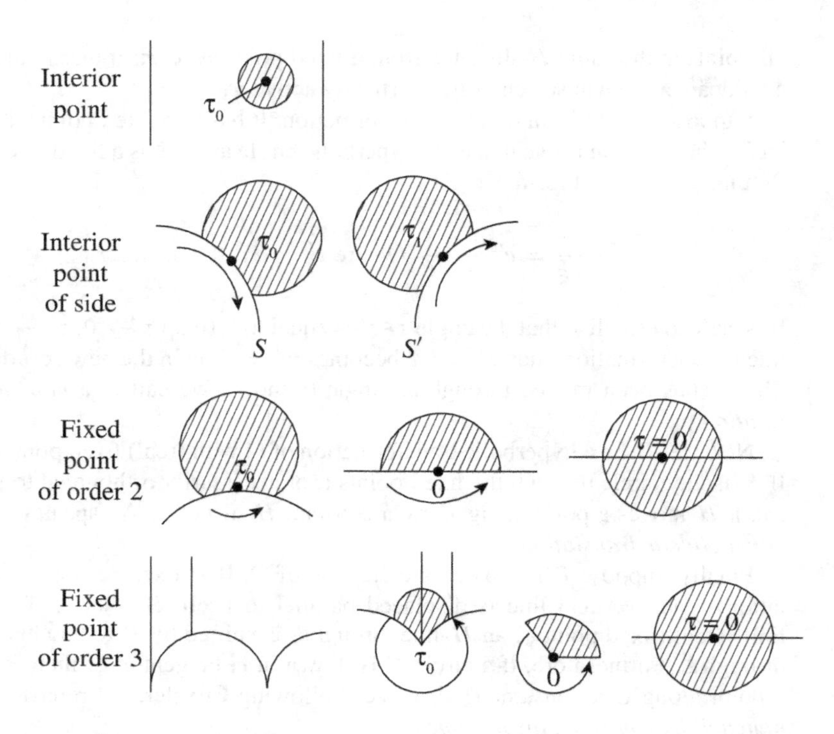

Now R'_G has 4 classes of points:

(1) τ_0 is an inner point of R_G. Let S be an open disk about τ_0 lying in R_G. Then $t = \tau - \tau_0$ maps S topologically onto a disk about $t = 0$ in the t-plane; S is an open neighborhood of τ_0 and the map $\tau \to t$ is a "disk homoeomorphism" (t is a "local variable" at τ_0).

(2) τ_0 is an interior point of a bounding arc s of R_G. There is an equivalent point τ_1 on an equivalent arc s'. Let S be a small open noneuclidean disk about τ_0 and $S' = \gamma S$ the equivalent disk about τ_1. A neighborhood N of τ_0 is $(S \cup S') \cap R_g$ together with the arcs of s and s' lying is S and S' respectively – cf. figure, p. 17. (Note that these arcs are identified.) N therefore consists of two "half-disks," N_0 and N_1. The linear transformation $\tau \to i(\tau - \tau_0)/(\tau - \bar{\tau}_0)$ sends N_0 onto D_0, the upper half of an ordinary disk D about the origin, while $\tau \to -i(\tau - \tau_1')/(\tau - \bar{\tau}_1')$ sends N_1 onto D_1, the lower half of D. Since the diameters of D_0 and D_1 are identified in the mapping, the two maps can be fitted together to give a homeomorphism of N onto D.

(3) τ_0 is a fixed point of G of order two, i.e., an image of i. Now τ_0 lies on a boundary arc s of R_G, so we take for its neighborhood a "half-disk" $N = (S \cap R_G) \cup (s \cap S)$, where S is a small noneuclidean open disk about τ_0; the two boundary arcs of N separated by τ_0 are identified. The linear transformation $\tau \to i(\tau - \tau_0)/(\tau - \bar{\tau}_0)$ maps N on the upper half of an ordinary disk about the origin with the bounding radii identified (see figure); hence

$t = (i(\tau - \tau_0)/(\tau - \bar{\tau}_0))^2$ maps N homeomorphically onto an open disk about $t = 0$; t is a local variable at τ_0.

(4) Finally, let τ_0 be a fixed point of order 3 (an image of $\rho = e^{2\pi i/3}$); τ_0 lies at the intersection of two bounding arcs of R_G. If one of these is a vertical side, we shift R_G to the left slightly; then τ_0 will appear as in the figure. In the neighborhood of τ_0, R_G subtends an angle $2\pi/3$ and the bounding arcs issuing from τ_0 are G-equivalent. Hence the neighborhood of τ_0 will be constructed in the same way as in 3), but the local variable is now $t = (i(\tau - \tau_0)/(\tau - \bar{\tau}_0))^3$.

Observe that the parabolic vertices (for example, i^∞) do not appear in R'_G. Hence in the Riemann surface $R = \mathbf{H}/G$ these will be projected into "punctures." There are as many punctures as there are distinct parabolic classes in G.

We can also think of $\mathbf{H}$ as a *covering surface* of the Riemann surface $R = \mathbf{H}/\Gamma$. If R is a Riemann surface, the surface $\tilde{R} = (\tilde{R}, \pi)$ is called a smooth covering surface of R provided the "projection mapping" $\pi : \tilde{R} \to R$ is a local homeomorphism. The points in the inverse image $\pi^{-1}(q), q \in R$, are said to lie over q. The surface $\tilde{R}$ is said to be unlimited if, given a curve γ on R with initial point q and a $\tilde{q} \in \tilde{R}$ lying over q, there is a curve $\tilde{\gamma}$ on $\tilde{R}$ with initial point $\tilde{q}$ that lies over γ. We also allow certain points in $\tilde{R}$ (branch points) where the one-to-one character of π is violated; if such points exist we say $\tilde{R}$ is a branched covering surface of R. The branch points are isolated and if $\tilde{B}$ is the set of branch points, $\tilde{R} - \tilde{B}$ is a smooth covering of $R - \pi(\tilde{B})$.

Returning to G and $\mathbf{H}$ we see that $(\mathbf{H}, \sigma)$ is a branched unlimited covering of $R = \mathbf{H}/G$, where the projection map is $\sigma : \mathbf{H} \to R$, the "identification map" referred to earlier. The branch points are the points of $\mathbf{H}$ fixed by elements of G (and so are necessarily images of i and ρ under the modular group). Analytic maps between surfaces can be defined and it can be shown that σ is analytic except at the branch points.

The process of projecting $\mathbf{H}$ *on* $\mathbf{H}/G$ can be reversed. Let R be a closed (compact) Riemann surface. Among all the smooth unlimited covering surfaces of R there is essentially only one that is simply connected; this surface covers all others and is called the *universal covering surface* $\hat{R} = (\hat{R}, \pi)$. Since $\hat{R}$ is simply connected, we can map it conformally and $1-1$ onto a normal plane region, either the extended plane, the finite plane, or the upper half-plane $\mathbf{H}$ (Riemann's Theorem). When R is of genus > 1, this map f is always onto $\mathbf{H}$ (hyperbolic case).

Now let h be a *covering transformation* of $\hat{R}$, i.e., h is a homeomorphism of $\hat{R}$ that carries a point lying over $q \in R$ into another point of $\hat{R}$ lying over q. In symbols, $\pi \circ h = \pi$. The map $V = f \circ h \circ f^{-1}$ is from $\mathbf{H}$ to $\mathbf{H}$ and is $1-1$ and analytic, since the analytic structure of $\hat{R}$ (induced from R) makes h analytic. Thus V is a real linear-fractional transformation, and the group of covering transformations of R is conjugate by f to a group $G \subset \Omega$. Besides this, it is easy to show G is discrete, so G is now a real discrete group. Furthermore, G is isomorphic to the fundamental group of R. Thus, given a closed Riemann surface R we have found a real discrete group acting on $\mathbf{H}$ such that $\mathbf{H}/G = R$. The group G is not unique, but a group G' satisfies $\mathbf{H}/G' = R$ if and only if $G' = xGx^{-1}$

with $x \in \Omega$. The theorem we have just asserted is one of the deepest of geometric function theory and is often called the Uniformization Principle.

As a consequence of the compactness of R, the fundamental region of G is relatively compact in $\mathbf{H}$. Hence G has no parabolic elements and so is not a subgroup of Γ (of finite index). By starting with a surface R which is closed except for a finite number of punctures and applying a procedure similar to the above, we do get groups G with parabolic elements. To get Γ itself, we must introduce a *branched* universal covering surface. In this case R will be a surface of genus 0 with one puncture and $\hat{R}$ will be branched over two points of R. The puncture will be the image of i^∞ in $\mathbf{H}$ under $\pi \circ f^{-1}$, and the two points in R images of i and ρ.

3. Let $\tau' = V\tau = x' + iy'$, $\tau = x + iy$, where $V = (\cdot\cdot|cd) \in \Omega$. For the area elements we have

$$dx'dy' = J\,dx\,dy,$$

with

$$J = \begin{vmatrix} \dfrac{\partial x'}{\partial x} & \dfrac{\partial x'}{\partial y} \\ \dfrac{\partial y'}{\partial x} & \dfrac{\partial y'}{\partial y} \end{vmatrix}.$$

By the Cauchy-Riemann equations,

$$J = \begin{vmatrix} \dfrac{\partial x'}{\partial x} & -\dfrac{\partial y'}{\partial x} \\ \dfrac{\partial y'}{\partial x} & \dfrac{\partial x'}{\partial x} \end{vmatrix} = \left(\frac{\partial x'}{\partial x}\right)^2 + \left(\frac{\partial y'}{\partial x}\right)^2 = \left|\frac{\partial \tau'}{\partial \tau}\right|^2,$$

so that

$$\frac{dx'dy'}{|d\tau'|^2} = \frac{dx\,dy}{|d\tau|^2}.$$

Multiplying this equation by $|d\tau'|^2/y'^2 = |d\tau|^2/y^2$ (cf. Note 1) we get the desired result.

4. Let $V = (a\,b|c\,d)$, $V' = (a'\,b'|c'\,d')$ with $V, V' \in GL^+(2, Z)$ and $a/c = a'/c' = r$. Then $V^{-1}V'$ fixes ∞ and so

$$V^{-1}V' = D = \begin{pmatrix} \alpha & \beta \\ 0 & \delta \end{pmatrix}.$$

where D has rational coefficients and determinant 1; note $\alpha/\delta > 0$. Hence

$$f|V' = f|VD = (f|V)\left(\frac{\alpha\tau + \beta}{\delta}\right).$$

which has the same limit for $\tau \to i^\infty$ as $f|V$.

5. Take the formula

$$\sum_{d=-\infty}^{\infty} (u+d)^{-2} = \pi^2 \operatorname{cosec}^2 \pi u = -4\pi^2 (e^{\pi i u} - e^{-\pi i u})^2$$

$$= \begin{cases} -4\pi^2 \sum_{m=1}^{\infty} m e^{2\pi i m u}, \operatorname{Im} u > 0 \\ -4\pi^2 \sum_{m=1}^{\infty} m e^{-2\pi i m u}, \operatorname{Im} u < 0 \end{cases}$$

and differentiate it $k-2$ times. In the result replace u by $c\tau$ and sum over c, using the top formula for $c > 0$ and the bottom one for $c < 0$. We get

$$(k-1)!(2\pi i)^{-k} \sum_{c \neq 0} \sum_{d=-\infty}^{\infty} (c\tau + d)^{-k}$$

$$= 2 \sum_{m=1}^{\infty} x^m \sum_{d|m} d^{k-1}.$$

Adding on the term for $c = 0$ yields

$$\sum_{\substack{c,d=-\infty \\ (c,d)\neq(0,0)}}^{\infty} (c\tau + d)^{-k} = 2\zeta(k) + \frac{2(2\pi)^k(-1)^{k/2}}{(k-1)!} \sum_{n=1}^{\infty} \sigma_{k-1}(n) x^n.$$

Now the left member equals

$$\sum_{\substack{\delta=1 \\ (c,d)=\delta}}^{\infty} = \sum_{\delta=1}^{\infty} \delta^{-k} \sum_{(c,d)=1} (c\tau + d)^{-k} = \zeta(k) G_k,$$

and this gives (25).

6. Define

$$P(\tau) = x \prod_{1}^{\infty} (1 - x^m)^{24}, \; x = e^{2\pi i \tau}.$$

C. L. Siegel has shown (Mathematika, 1954, or ch. V Notes, No. 3) that $P \in \{\Gamma, -12\}$. Hence $f = \Delta / P \in \{\Gamma, 0\}$ and f is regular in $\mathbf{H}$ (because, as an infinite product, P is never 0). Moreover, at i^{∞} we have $f = 1 + a_1 x + \dots$. Thus f is bounded in the fundamental region. Later we shall prove (theorem 3) that a bounded modular function is a constant. Hence $P = \Delta$.

Chapter II

Modular Forms with Multiplicative Fourier Coefficients. I.

1 Introduction

In previous lectures we have met the arithmetic functions $\sigma_{k-1}(n)$ and $\tau(n)$, occurring as the Fourier coefficients of the modular forms E_k and Δ, respectively. They have the following properties:

$$f(mn) = f(m)f(n) \text{ for } (m,n) = 1 \tag{1}$$
$$f(p^{\alpha+1}) = f(p)f(p^{\alpha}) - p^{k-1}f(p^{\alpha-1}), \alpha = 1, 2, \ldots \tag{2}$$

where we make the convention that

$$f(x) = 0 \tag{2a}$$

if f is an arithmetic function and x is not an integer. The function $\tau(n)$ satisfies (1) and (2) with $k = 11$. These equations reduce the calculation of $\sigma_{k-1}(n)$ or $\tau(n)$ to their calculation for $n =$ prime. We shall call any arithmetic function, not identically zero, which satisfies (1) and (2), *multiplicative*, but it must be remembered that the class of functions thus defined is more restrictive than what is usually called multiplicative in number theory, where only property (1) is assumed.

The multiplicativity of $\tau(n)$ is not trivial; it was conjectured by Ramanujan and first proved by Mordell. In 1936 Hecke developed a wide generalization: *he found all modular forms* – Δ is only one of them – *that have multiplicative Fourier coefficients.* This was done first for Γ and later for $\Gamma(n)$, the principal congruence subgroup modulo n (Math. Ann. 114 (1937), 1–28; 316–335). This extensive theory was, however, deficient in one crucial point, which was later supplied by Petersson (Math. Ann. 116 (1939), 401–412; 117 (1939), 39–64).

Hecke's basic idea was to introduce certain linear operators T_n, $n = 1, 2, \ldots$, each of which maps the space of modular forms of a given degree into itself. In addition these operators were chosen so that the multiplicativity properties of the Fourier coefficients of a form f correspond exactly to the property that f shall be an eigenfunction of all T_n (a *simultaneous eigenfunction*) and that its first Fourier coefficient should be 1. However, Hecke did not succeed in proving the existence of enough simultaneous eigenfunctions. This was done by Petersson, who defined for this purpose a scalar product (f, g) of forms in $\langle\Gamma\rangle_0$, making it a *Hilbert space*. He proved, moreover, that the T_n were *hermitian* operators on $\langle\Gamma\rangle_0$ and was able then, by a known theorem of algebra, to simultaneously diagonalize the matrices representing the T_n. This provided the final solution of the problem, which is to the effect that $\langle\Gamma\rangle_0$ admits an orthogonal basis of simultaneous eigenfunctions each of which has multiplicative Fourier coefficients, and that this system is unique.

We shall now present the proofs of the foregoing developments.

2 The Operators T_p

Let k be a fixed even interger. For abbreviation we write $\langle\Gamma\rangle$ for $\langle\Gamma, -k\rangle$, similarly $\langle\Gamma\rangle_0$ for $\langle\Gamma, -k\rangle_0$, these denoting the vector space of holomorphic modular forms that are finite and that vanish, respectively, at the cusp i^∞. We recall from Chapter I, (28) that

$$\mu = \dim\langle\Gamma\rangle_0 = \begin{cases} [k/12], k \not\equiv 2 \pmod{12} \\ [k/12] - 1, k \equiv 2 \pmod{12}, k > 2 \\ 0, k = 2 \end{cases} \qquad (3)$$

We obtain linear operators that preserve $\langle\Gamma\rangle_0$ as follows. Let $H_p \equiv H$ be any set of 2×2 integral matrices of positive determinant such that in

$$H = \Gamma \cdot B,$$

B is a finite set. Let $f \in \langle\Gamma\rangle_0$. Then we define a linear operator X by

$$f|X = \sum_{M \in B} f|M.$$

Note that $f|X$ is independent of the choice of B. Indeed, if $H = \Gamma \cdot B'$, then each $b_i \in B$ is associated with a $b_i' = \gamma_i b_i$, $\gamma_i \in \Gamma$ that is in B', and $f|b_i' = (f|\gamma_i)|b_i = f|b_i$.

Now let $V \in \Gamma$; then

$$(f|X)|V = \sum_{M \in B} f|MV = f|X,$$

since clearly $H = \Gamma \cdot BV$. Because f is a cusp form, $(f|M)_\infty = 0$ and so we have proved that $f|X \in \langle\Gamma\rangle_0$. Hence

$$\langle\Gamma\rangle_0|X \subset \langle\Gamma\rangle_0. \tag{4}$$

There is a wide choice of operators X satisfying the required condition, but we select one that is connected with the problem of multiplicativity. Let

$$H_p = \left\{ \begin{pmatrix} a & b \\ c & d \end{pmatrix} \middle| \, a, b, c, d \in Z, ad - bc = p \right\}, \, p = \text{prime.} \tag{5}$$

Lemma 1. *$H_p = \Gamma \cdot B_p$ with*

$$B_p = \left\{ M_j, 0 \leqq j \leqq p; M_j = \begin{pmatrix} 1 & j \\ 0 & p \end{pmatrix}, 0 \leqq j < p, M_p = \begin{pmatrix} p & 0 \\ 0 & 1 \end{pmatrix} \right\}.$$

Proof. Let $(a\, b|c\, d) \in H_p$. Now

$$\begin{pmatrix} A & B \\ C & D \end{pmatrix} \begin{pmatrix} a & b \\ c & d \end{pmatrix} = \begin{pmatrix} \cdot & \cdot \\ 0 & \cdot \end{pmatrix}$$

if $Ca + Dc = 0$, so choose $C = -c/t$, $D = a/t$, where $t = (a, c)$. Then $(C, D) = 1$ and we can find A, B so that $(A\, B|C\, D) \in \Gamma$. Hence

$$\begin{pmatrix} A & B \\ C & D \end{pmatrix} \begin{pmatrix} a & b \\ c & d \end{pmatrix} = \begin{pmatrix} 1 & l \\ 0 & p \end{pmatrix} \text{ or } \begin{pmatrix} p & m \\ 0 & 1 \end{pmatrix}.$$

In the first case choose x so that $0 \leqq l + xp < p$; then

$$\begin{pmatrix} 1 & x \\ 0 & 1 \end{pmatrix} \begin{pmatrix} 1 & l \\ 0 & p \end{pmatrix} = \begin{pmatrix} 1 & j \\ 0 & p \end{pmatrix}, 0 \leqq j < p$$

and

$$\begin{pmatrix} a & b \\ c & d \end{pmatrix} = V \begin{pmatrix} 1 & j \\ 0 & p \end{pmatrix} \text{ with } V \in \Gamma.$$

In the second case note that

$$\begin{pmatrix} 1 & -m \\ 0 & 1 \end{pmatrix} \begin{pmatrix} p & m \\ 0 & 1 \end{pmatrix} = \begin{pmatrix} p & 0 \\ 0 & 1 \end{pmatrix}, \text{ or } \begin{pmatrix} a & b \\ c & d \end{pmatrix} = W \begin{pmatrix} p & 0 \\ 0 & 1 \end{pmatrix}, W \in \Gamma. \qquad \text{Q.E.D.}$$

It is convenient to normalize the definition of the operator defined by H_p; we set

$$f|T_p = p^{k-1} \sum_{M \in B_p} f|M. \tag{6}$$

From the discussion preceding the lemma we can now assert the second statement of

Theorem 1.

$$\langle\Gamma\rangle|T_p \subset \langle\Gamma\rangle,$$
$$\langle\Gamma\rangle_0|T_p \subset \langle\Gamma\rangle_0.$$

The first statement is proved in the same way.

Explicitly we have

$$f|T_p = \frac{1}{p}\sum_{j=0}^{p-1} f\left(\frac{\tau+j}{p}\right) + p^{k-1} f(p\tau). \tag{6a}$$

The connection of the operators $\{T_p\}$ with the problem of multiplicativity is made clear by

Theorem 2. *If f has the Fourier series $\sum_{m \geqq 8} a(m)x^m$, where $s \geq 0$, then*

$$f|T_p = \sum_{m \geqq 8/p} \left[a(mp) + p^{k-1} a\left(\frac{m}{p}\right)\right] x^m. \tag{7}$$

Here $x = e^{2\pi i\tau}$ as usual and, by our convention (2a), $a(m/p) = 0$ when $p \nmid m$. Setting $a(l) = 0$ for $l < s$, we have

$$\begin{aligned} f|T_p &= p^{-1} \sum_{m \geq 0} a(m) e^{2\pi i \frac{m\tau}{p}} \sum_{j=0}^{p-1} e^{2\pi i \frac{mj}{p}} + p^{k-1} \sum_{m \geq 0} a(m) e^{2\pi i m p \tau} \\ &= \sum_{m \geq 0} a(mp) x^m + p^{k-1} \sum_{m \geq 0} a(m/p) x^m. \end{aligned}$$

In the first sum the range of m can be taken to be $m \geq s/p$, and this is also true of the second sum, since $a(l) = 0$ for $0 \leq m/p = l < s/p$.

Next we observe that the T_p commute. In fact note first that we can write

$$f|T_p = p^{k-1} \sum_{\substack{nd=p, d>0 \\ b \bmod d}} f|\begin{pmatrix} a & b \\ 0 & d \end{pmatrix}.$$

because of $f(\tau+1) = f(\tau)$. Thus for $p \neq p'$,

$$f|T_p T_{p'} = (pp')^{k-1} \sum_{a,d} \sum_{a'd'} (dd')^k f| \begin{pmatrix} a & b \\ 0 & d \end{pmatrix} \begin{pmatrix} a' & b' \\ 0 & d' \end{pmatrix}.$$

and the product of the matrices is

$$\begin{pmatrix} aa' & ab'+bd' \\ 0 & dd' \end{pmatrix}.$$

Now as b runs mod d and b' mod d', $ab'+bd'$ runs mod dd', since $(a, d') = 1$ because of $a|p, d'|p'$. Hence

$$f|T_pT_{p'} = (pp')^{k-1} \sum_{\substack{\alpha\delta=pp',\delta>0 \\ \beta \bmod \delta}} (\delta\delta')^k f| \begin{pmatrix} \alpha & \beta \\ 0 & \delta \end{pmatrix} = f|T_{p'}T_p,$$

so we have proved

Theorem 3. *The operators in the set $\{T_p, p = \text{prime}\}$ commute.*

Note that the only invariance property of f used in the proof was $f(\tau+1) = f(\tau)$.

3 Simultaneous Eigenforms

The crucial role of an eigenfunction of T_p is not far to seek. Let f be an eigenfunction, i.e.,

$$f|T_p = \lambda(p)f.$$

Inserting the Fourier series and using theorem 2, we get, on equating coefficients,

$$a(mp) + p^{k-1}a(m/p) = \lambda(p)a(m).$$

Let us suppose not only that $a(1) \neq 0$ but

$$a(1) = 1,$$

in which case we say that the eigenfunction is *normalized*. Then putting $m = 1$ in the above equation we see that

$$\lambda(p) = a(p), \tag{8}$$

and so

$$a(mp) + p^{k-1}a(m/p) = a(m)a(p). \tag{9}$$

In particular, for $m = p^\alpha$:

$$a(p^{\alpha+1}) = a(p)a(p^\alpha) - p^{k-1}a(p^{\alpha-1}), \quad \alpha \geqq 0. \tag{10}$$

Instead of the term "eigenfunction" we shall use "eigenform", to remind us that f is a modular form.

Now suppose f is a normalized eigenform of *all* T_p, or what we shall call a *normalized simultaneous eigenform*. Equation (10) is the multiplicativity property (2), and (9) and (10) together imply (1): $a(mn) = a(m)a(n)$ for $(m, n) = 1$. In fact (1) is true for arbitrary m and $n = p^\alpha$, $\alpha = 1$; this is simply (9), as $p \nmid m$ (hence $a(m/p) = 0$). Assuming it true for $\alpha < \beta$, we take $m = np^{\beta-1}$ in (9) and get

$$a(p^\beta n) = a(p)a(p^{\beta-1})a(n) - p^{k-1}a(p^{\beta-2})a(n) = a(p^\beta)a(n),$$

where we used (10) with α replaced by $\beta - 1$. By induction[1] on the number of primes dividing m, we establish (1). Hence we have proved the first part of

Theorem 4. *If $f \in \langle\Gamma\rangle_0$, a normalized simultaneous eigenform of all T_p, $p =$ prime, and $f(\tau) = \sum_{m=1}^{\infty} a(m)x^m$, $a(1) = 1$, then $a(n)$ is multiplicative in the sense of (1) and (2). Conversely, if $f \in \langle\Gamma\rangle_0$ has the above Fourier series whose coefficients satisfy (1), (2), and $a(1) = 1$, then f is a normalized simultaneous eigenform of all T_p with eigenvalue $a(p)$.*

To prove the converse we first show easily that (1) and (2) imply (9). Then by theorem 2 and (9),

$$f|T_p = \sum_{m=1}^{\infty}[a(mp) + p^{k-1}a(m/p)]x^m = a(p)\sum_{m=1}^{\infty} a(m)x^m.$$

Since our basic goal is to find modular forms with multiplicative coefficients, theorem 4 shows we must prove the existence of normalized simultaneous eigenforms of all $\{T_p\}$.

4 Simultaneous Diagonalization

The essential fact is now that the vector space $\langle\Gamma\rangle_0$ is *finite-dimensional*; the dimension μ appears in (3). The operator T_p can as a consequence be represented by a $\mu \times \mu$ matrix with respect to a basis $\{f_1, \ldots, f_\mu\}$ of $\langle\Gamma\rangle_0$:

$$f_j|T_p = \sum_{i=1}^{\mu} \lambda_{ij}(p) f_i, \quad j = 1, 2, \ldots, \mu. \tag{11}$$

The commutativity of the $\{T_p\}$ induces commutativity in the matrices $\{\lambda(p)\} = \{\lambda_{ij}(p)\}$. Suppose in addition we can show that

$$\lambda(p) \text{ is } \textit{hermitian } (\lambda_{ij}(p) = \bar{\lambda}_{ji}(p)) \text{ for all } p. \tag{12}$$

Among the matrices $\{\lambda(p)\}$ there are a finite number ($\leqq \mu^2$) that are linearly independent, say $\lambda(p_1), \ldots, \lambda(p_n)$. By a theorem of algebra* the set $\{\lambda(p_i)\}$ can be

*Gantmakher, F. R., The Theory of Matrices, vol. 1, ch. IX. sec. 15, Chelsea Publ. Co., N.Y. 1959; or Newman, Morris, *J. Res. Nat. Bur. Stand.* (U.S.), 71B (*Math. Sci.*), Nos. 2 and 3, 69–71 (Apr.–Sept. 1967).

simultaneously diagonalized by a unitary matrix. Since each $\lambda(p)$ is a finite linear combination of the $\{\lambda(p_i)\}$, it follows that the matrix A diagonalizing the $\{\lambda(p_i)\}$ also diagonalizes all $\{\lambda(p)\}$. Let $\{g_1, \ldots, g_\mu\}$ be the basis obtained by applying A to $\{f_i\}$; it is a basis, as A is unitary. In the new basis T_p is represented by the *diagonal* matrix $A^{-1}\lambda(p)A$. Hence $\{g_i\}$ is a basis of simultaneous eigenforms of all $\{T_p\}$.

It was Petersson who proved the hermitian character of the $\{\lambda(p)\}$. He did this by defining a scalar product[2] in $\langle\Gamma\rangle_0$, thereby making it a *Hilbert space*. For $f, g \in \langle\Gamma\rangle_0$ define

$$(f, g; R_\Gamma) = \iint\limits_{\dot{R}_\Gamma} f(\tau)\bar{g}(\tau)v^{k-2}du\,dv, \ \tau = u + iv. \tag{13}$$

It is proved in Short Course, p. 98, that the integral converges and is independent of the choice of[3] R_Γ. (Although the proof given there assumes $k > 2$, it is actually valid for $k = 2$.) Hence we can write simply (f, g). That the integral defines a scalar product is clear. With these properties $\langle\Gamma\rangle_0$, together with the scalar product (f, g), becomes a Hilbert space – still called $\langle\Gamma\rangle_0$ – since the space is finite-dimensional and so necessarily complete.

We can assume from the beginning that the basis $\{f_1, \ldots, f_\mu\}$ is an *orthonormal* basis:

$$(f_i, f_j) = \delta_{ij}.$$

Then (12) is equivalent to

$$(f_i|T_p, f_j) = (f_i, f_j|T_p), \quad i, j = 1, \ldots, \mu, \tag{14}$$

as is trivially verified. Hence we must prove (14), which in turn is equivalent to

$$(f|T_p, g) = (f, g|T_p) \text{ for } f, g \in \langle\Gamma\rangle_0. \tag{15}$$

The proof of (15) is rather lengthy and we shall not give it here; it can be found in the first paper of Petersson cited at the beginning of this chapter. In essence the ideas of the proof are as follows. The scalar product $(f|T_p, g)$ may be expressed as a sum, of which a typical term is[4]

$$\iint\limits_{R_\Gamma} (f(\tau)|M_j)\bar{g}(\tau)v^{k-2}du\,dv = p^{-k} \iint\limits_{M_j R_\Gamma} f(\tau)(\bar{g}(\tau)|M_j^{-1})v^{k-2}du\,dv.$$

This begins to look like what we want except that $M_j R_\Gamma$ is quite different for different j. Therefore we go to the subgroup $\Gamma(p)$, whose normality will help us. Write $\Gamma = \Gamma(p) \cdot A$ and note that $f|A_i = f$ for $A_i \in A$; clearly

$$\iint\limits_{R_{\Gamma(p)}} f\bar{g}v^{k-2}du\,dv = [\Gamma : \Gamma(p)] \cdot \iint\limits_{R_\Gamma} f\bar{g}v^{k-2}du\,dv.$$

We may therefore prove the result for integrals extended over fundamental regions $R_p = R_{\Gamma(p)}$ consisting of $[\Gamma : \Gamma(p)]$ copies of R_Γ. Now we take advantage of the fact that $X^{-1}\Gamma(p)X \subset \Gamma$ for any $X \in H_p$, which leads to the lemma that $(f, g; R_p) = (f, g; X^{-1}R_p)$. Then we can make the above change of variable and get

$$\iint_{R_{(p)}} (f(\tau)|M_j)\bar{g}(\tau)v^{k-2}du\,dv = \iint_{M_j^{-1}R_{(p)}} \cdots$$

$$= p^{-k} \iint_{R_p} f(\tau)(\bar{g}(\tau)|M_j^{-1})v^{k-2}du\,dv. \qquad (*)$$

Now it is possible to select a system B_p such that not only is $H_p = \Gamma \cdot B_p$ but $H_p = B_p \cdot \Gamma$. For example:

$$B_p = \{S_j, 0 \leqq j \leqq p\}, S_j = \begin{pmatrix} 1 & j \\ j & j^2 + p \end{pmatrix}, 0 \leqq j < p;\ S_p = \begin{pmatrix} p & 0 \\ 0 & 1 \end{pmatrix}.$$

And of course $H_p = pH_p^{-1} = \Gamma \cdot B_p \Rightarrow H_p = pB_p^{-1} \cdot \Gamma$, so $f|T_p = p^{k-1} \cdot p^{-k} \sum_j f|M_j^{-1}$. Hence summing (*) over j, we get the desired result.

We now return to the orthonormal basis $\{f_1, \ldots, f_\mu\}$ of $\langle\Gamma\rangle_0$. Let A be the unitary matrix diagonalizing all $\lambda(p)$, as above. The new basis $\{g_1, \ldots, g_\mu\}$ obtained by applying A to $\{f_i\}$ is also orthonormal; write the Fourier series

$$g_i = \sum_{m=1}^{\infty} b_i(m)x^m, i = 1, 2, \ldots, \mu.$$

According to what we have said above, each g_i is a simultaneous eigenform of all $\{T_p\}$ and we write

$$g_i|T_p = \Lambda_i(p)g_i.$$

Let us consider any g_i and suppose $b_i(r)$, $r \geq 1$, is its first nonzero coefficient. If $r > 1$, let p be a prime dividing r. By (7),

$$g_i|T_p = b_i(r)x^{r/p} + \ldots = \Lambda_i(p)b_i(r)x^r + \ldots,$$

which implies $b_i(r) = 0$, a contradiction. Hence there is no such prime p and r must be 1. That is, all g_i begin with a term in x. Therefore we can normalize by setting $h_i = g_i/b_i(1)$, $i = 1, \ldots, \mu$; then

$$h_i(\tau) = \sum_{m=1}^{\infty} c_i(m)x^m, c_i(1) = 1.$$

The new basis $\{h_i\}$ is still orthogonal but no longer orthonormal. Since the $\{h_i\}$ are normalized simultaneous eigenforms of all $\{T_p\}$, their Fourier coefficients are multiplicative, by theorem 4. We have therefore proved:

Theorem 5. *For every even positive integer k the Hilbert space $\langle\Gamma, -k\rangle_0$ admits an orthogonal basis of forms*

$$h_i(\tau) = \sum_{m=1}^{\infty} c_i(m)x^m, c_i(1) = 1 \tag{16}$$

where, for each i, the Fourier coefficients $c_i(m)$ have the multiplicative properties (1) and (2). Each h_i is a normalized simultaneous eigenform of all $\{T_p, p = \text{prime}\}$ with eigenvalue $\Lambda_i(p)$ and

$$\Lambda_i(p) = c_i(p). \tag{17}$$

So far we have T_p only for prime p; the reason was simplification of the proofs. However, we can now extend T_p to T_n for all $n = 1, 2, \ldots$, by defining*

$$\begin{aligned} T_i &= \text{identity operator,} \\ T_{p^{\alpha+1}} &= T_p T_{p^\alpha} - p^{k-1} T_{p^{\alpha-1}}, \alpha \geqq 1 \\ T_n &= \prod_{p^e \| n} T_{p^e}; \end{aligned}$$

that is, we make T_n satisfy (1) and (2). Then it is shown without difficulty that the full set $\{T_n, n = 1, 2, \ldots\}$ has the same properties as $\{T_p\}$: they commute and are hermitian. The $\{h_i\}$ are now normalized simultaneous eigenforms for all $\{T_n\}$ and (17) is satisfied for all n:

$$\Lambda_i(n) = c_i(n), n = 1, 2, \ldots; i = 1, 2, \ldots, \mu. \tag{18}$$

As a generalization of (1) it can be shown by induction that

$$c(m)c(n) = \sum_{d|(m,n)} c(mn|d^2) \cdot d^{k-1}$$

for all $m, n = 1, 2, \ldots$. The same relation is then satisfied by the $\{T_n\}$.

5 Uniqueness

The final step is to prove the uniqueness of the normalized simultaneous eigenforms $h_1, h_2, \ldots, h_\mu$ (apart from order). For later purposes we shall prove a more general theorem.

Theorem 6. *Let $g \in \langle\Gamma\rangle_0$, $g \not\equiv 0$, be a simultaneous eigenform for all T_p with prime $p \neq q$, where q is an arbitrary fixed prime. Then*

$$g(\tau) = ch_i(\tau), c \neq 0$$

for some i in $1 \leqq i \leqq \mu$.

* $p^e \| n$ means $p^e | n$, $p^{e+1} \nmid n$.

Proof. We first prove: for each pair $i \neq j$ in $1 \leqq i, j \leqq \mu$ there is an $n > 1$, $(n, q) = 1$, such that

$$\Lambda_i(n) \neq \Lambda_j(n). \tag{19}$$

if not, we can write, by (18),

$$h_i(\tau) - h_j(\tau) = \varphi(q\tau).$$

where $\varphi(\tau) = \sum_{n=1}^{\infty} a(n) e^{2\pi i n\tau}$. Plainly $\varphi(q\tau) \in \langle\Gamma\rangle_0$. We now need two lemmas. Q.E.D.

Lemma 2. *If $\psi(\tau)|S = \psi(\tau)$, $S = (1\,1|0\,1)$, and $\psi(n\tau) \in \langle\Gamma_0(n)\rangle_0$, $n > 1$, then $\psi(\tau) \in \langle\Gamma\rangle_0$.*

First, $\psi(\tau) \in \Gamma^0(n)$. For let $\omega(\tau) = \psi(n\tau)$; then $\omega(\tau) \in \langle\Gamma_0(n)\rangle$. Hence

$$\begin{aligned}
\psi(\tau)\left|\begin{pmatrix} a & nb \\ c & d \end{pmatrix}\right. &= (c\tau + d)^{-k} \psi\left(\frac{a\tau + nb}{c\tau + d}\right) \\
&= (c\tau + d)^{-k} \omega\left(\frac{a\left(\frac{\tau}{n}\right) + b}{nc\left(\frac{\tau}{n}\right) + d}\right) = \left\{\omega(z)\left|\begin{pmatrix} a & b \\ nc & d \end{pmatrix}\right.\right\}_{z=\tau/n} \\
&= \omega\left(\frac{\tau}{n}\right) = \psi(\tau),
\end{aligned}$$

Since

$$\begin{pmatrix} a & b \\ nc & d \end{pmatrix} \in \Gamma_0(n) \text{ if } \begin{pmatrix} a & nb \\ c & d \end{pmatrix} \in \Gamma^0(n).$$

But $\psi|S = \psi$; hence with $T = (0\,-1|1\,0)$ we get: $\psi|TS = \psi|STS = \psi|(1\,0|1\,1) = \psi = \psi|S$, or $\psi|T = \psi$. Thus $\psi|S = \psi$, $\psi|T = \psi$ so $\psi \in \{\Gamma, -k\}$. Since $\psi(n\tau) \in \langle\Gamma_0(n)\rangle_0$, we have $\lim_{\tau\to\infty} \psi(n\tau) = 0$, which implies $\lim_{\tau\to\infty} \psi(\tau) = 0$. Hence $\psi \in \langle\Gamma\rangle_0$.

Lemma 3. *If $\psi(\tau)$ and $\psi(n\tau)$ both belong to $\langle\Gamma\rangle_0$ for some $n > 1$, then $\psi(\tau) \equiv 0$.*

Proof. With $\psi(n\tau) = \omega(\tau) \in \langle\Gamma\rangle_0$, we get

$$\psi(n\tau) = \omega(\tau) = \omega(\tau)|T = \tau^{-k}\omega\left(-\frac{1}{\tau}\right) = \tau^{-k}\psi\left(\frac{-n}{\tau}\right).$$

or replacing τ by $n\tau$,

$$\psi(n^2\tau) = n^{-k}\tau^{-k}\omega\left(-\frac{1}{\tau}\right) = n^{-k}\psi(\tau)|T.$$

But $\psi(\tau) \in \langle\Gamma\rangle$, hence

$$\psi(n^2\tau) = n^{-k}\psi(\tau).$$

By inserting the Fourier series of ψ we see that this relation can hold only if all Fourier coefficients are zero, since $n > 1$.

We return to the proof of theorem 6 and choose $\varphi = \psi$, $n = q$. Recall that $\varphi(q\tau)$ is in $\langle\Gamma\rangle_0$, hence in $\langle\Gamma_0\rangle_0$. Then by lemma 2, $\varphi(\tau) \in \langle\Gamma\rangle_0$ and so by lemma 3, $\varphi \equiv 0$, $h_i = h_j$. Hence (19) is proved.

Now let $g(\tau)$ satisfy the hypotheses. Since $\{h_i, i = 1, \ldots, \mu\}$ is a basis for $\langle\Gamma\rangle_0$, we have

$$g = \sum_{j=1}^{\mu} \alpha_j h_j.$$

But $g \not\equiv 0$ and so some α_j is not zero. Suppose it is α_1; we shall show $\alpha_j = 0$ for $j > 1$.

Let $1 < l \leqq \mu$ and let T_m be such that

$$\Lambda_1(m) \neq \Lambda_l(m).$$

Call $\Lambda'(m)$ the eigenvalue of g with respect to T_m. Then $g|T_m = \Lambda'(m)g = \Lambda'(m) \sum_{j=1}^{\mu} \alpha_j h_j$. But $g|T_m = \left(\sum_{j=1}^{\mu} \alpha_j h_j\right) |T_m \sum_{j=1}^{\mu} \alpha_j \Lambda_j(m) h_j$. For $j = 1$: $\Lambda'(m)\alpha_1 = \alpha_1\Lambda_1(m)$, or $\Lambda'(m) = \Lambda_1(m)$, since $\alpha_1 \neq \bar{0}$. For $j = l$: $\Lambda'(m)\alpha_l = \alpha_l\Lambda_l(m)$ and so

$$0 = (\Lambda'(m) - \Lambda_l(m))\alpha_l = (\Lambda_1(m) - \Lambda_l(m))\alpha_l.$$

Since the factor of α_l is not 0, $\alpha_l = 0$ for $l > 1$, as promised. Q.E.D.

We can state the desired uniqueness theorem as an easy corollary of theorem 6.

Corollary. *The $\{h_i, i = 1, 2, \ldots, \mu\}$ are uniquely determined (apart from order) by the two properties:*

(i) the $\{h_i\}$ form a basis for $\langle\Gamma\rangle_0$.

(ii) each h_i is a normalized simultaneous eigenform of all T_p for prime p.

Chapter II Notes

1. Let $(m, n) = 1$ and suppose $m = p_1^{e_1} \cdots p_r^{e_r}$. By what was proved in the text we have

$$\begin{aligned} a(mn) &= a(p_1^{e_1} \cdot (p_2^{e_2} \cdots p_r^{e_r} n)) \\ &= a(p_1^{e_1})a(p_2^{e_2} \ldots p_r^{e_r} n)) = \ldots = a(p_1^{e_1}) \ldots a(p_r^{e_r})a(n). \end{aligned}$$

In particular this equation holds for $n = 1$, so

$$a(m) = a(p_1^{e_1}) \ldots a(p_r^{e_r})$$

and the result follows.

2. This scalar product is equivalent to the one used in Riemann surface theory. Let us first see how to integrate over a region S on a Riemann surface R. Let U

be an open neighborhood (parametric neighborhood) in S and $t = \Phi(q)$, $q \in U$, the corresponding homeomorphism; t is called a local variable for U. Suppose $\Delta = \Phi(\Delta')$, where Δ is a euclidean triangle; we call Δ' *a triangle on* R. If φ is an analytic function on S and $\varphi(q) = \varphi(\Phi^{-1}(t)) = \varphi_1(t)$ for $q \in U$, we define

$$\iint_{\Delta} \varphi = \iint_{\Delta} \varphi_1(t) dx\, dy, t = x + iy.$$

To make this definition invariant under changes of local variable we must require that $\varphi_1 dxdy$ be a *second-order differential* Ω in the following sense: to each local variable $t_\alpha = x_\alpha + idy_a$ there shall correspond an analytic function φ_α such that $\Omega = \varphi_\alpha dx_\alpha dy_\alpha$, and if t_β is another local variable, then

$$\varphi_\alpha(t_\alpha) dx_\alpha dy_\alpha = \varphi_\beta(t_\beta) dx_\beta dy_\beta. \qquad (*)$$

We now partition S into a network of triangles each of which is small enough to lie in one parametric neighborhood and define the integral over S to be the sum of the integrals over the triangles; it has to be shown, of course, that the result is independent of the triangulation used. Cf. R. Nevanlinna, Uniformisierung, Springer, Berlin (1953) 117–121.

Suppose φdt, ψdt are two ordinary differentials, i.e., φ, ψ are analytic functions of t and $\varphi(t)dt$, $\psi(t)dt$ are invariant under changes of the local variable t. Let $t' = x' + iy'$ be another local variable and let $\varphi(t)dt = \varphi'(t')dt'$, etc. The Jacobian of (x', y') with respect to (x, y) is $|dt'/dt|^2$. Hence

$$\begin{aligned}\varphi'(t')\bar{\psi}'(t')dx'dy' &= \varphi'(t')dt'\bar{\psi}'(t')d\bar{t}'\frac{dx'dy'}{dt'd\bar{t}'} = \varphi(t)dt\bar{\varphi}(t)d\bar{t}\cdot\frac{dt'}{dt}\frac{d\bar{t}'}{d\bar{t}}\frac{dx\,dy}{dt'd\bar{t}'}\\ &= \varphi(t)\bar{\psi}(t)dx\,dy,\end{aligned}$$

so that $\varphi\bar{\psi}dx\,dy$ satisfies (*) and is a second-order differential. Hence

$$\iint_R \varphi(t)\bar{\psi}(t)dx\,dy$$

is well-defined. This is defined to be the scalar product (φ, ψ) of two differentials φdt, ψdt. Cf. Nevanlinna, Uniformisierung, pp. 339–340.

Let us now return to the Petersson scalar product in the τ-plane and restrict our considerations to cusp forms f, g of degree -2 on the group G. We have

$$(f, g) = \int_{R_G}\int f(\tau)\bar{g}(\tau)du\,dv, \tau = u + iv.$$

Let $R = \mathbf{H}/G$; to the forms $fd\tau$, $gd\tau$ on $\mathbf{H}$ there correspond differentials φdt, ψdt on R. If σ is the mapping $\mathbf{H} \to R$ (the map that identifies G-equivalent points in $\mathbf{H}$), then σ^{-1} is single-valued and analytic when its range is restricted to R_G; hence so is $\tau = \sigma^{-1} \circ \Phi^{-1}(t)$. We now have

$$f(\tau)\bar{g}(\tau) du\, dv = f(\tau) d\tau \cdot \bar{g}(\tau) d\bar{\tau} \cdot du\, dv/d\tau\, d\bar{\tau}$$
$$= f(t)dt \cdot \bar{g}(t)d\bar{t} \cdot \left|\frac{d\tau}{dt}\right|^2 \cdot \frac{dx\, dy}{|d\tau|^2} = f(t)\bar{g}(t) dx\, dy.$$

Since $\sigma^{-1}(R) = R_G$, we have shown that σ is an *isometric* mapping of R_G onto R; that is, $(f, g) = (\varphi, \psi)$. The same conclusion holds for differentials of higher weight (corresponding to forms of higher negative degree), but the analogous scalar product has apparently not yet been used in the theory of Riemann surfaces.

3. The problem of convergence arises at the cusps of the fundamental region. Consider the cusp at i^∞. Since $f \in \langle\Gamma\rangle_0$ we have

$$f(\tau) = \sum_{m=1}^{\infty} a(m) e^{2\pi i m \tau} = O(e^{-2\pi y});$$

similarly for g. Thus the integrand is $O(y^{k-2} e^{-4\pi y})$ and this converges over any region $|x| \leqq A$, $y \geqq B$. (Note that this would be true even if only one of f, g is a cusp form provided the other is regular at i^∞.)

This disposes of the question for Γ. Now suppose G is a subgroup of finite index in Γ and let $p = A\infty$, $A \in \Gamma$, be a finite cusp. As we have seen (ch. I, sec. 3, (24b))

$$(\tau - r)^k f(\tau) = \sum_{m=1}^{\infty} b(m) e^{-2\pi i/c(\tau - r)}, c > 0$$

so that $f = 0(v^{-k} e^{-2\pi/cv})$ as $\tau = u + iv$ approaches $\tau = r$ from within a fundamental region. Hence the integrand is $0(v^{-k-2} e^{-4\pi/cv})$ and the integral is finite when extended over a triangular cusp region with apex at r. Since the index is finite, there are only a finite number of parabolic cusps in a fundamental region, and the convergence of the integral is proved.

Suppose $V \in G$; write $\tau = u + iv$, $\tau' = u' + iv'$. Recall that $du\, dv/v^2 = du'dv'/dv'^2$ and $|c\tau + d|^2 = v/v'$; also that c, d are real in $V = (a\, b | c\, d)$. Then

$$v'^k f(\tau')\bar{g}(\tau') \cdot du'dv'/v'^2 = f|V \cdot \bar{g}|V \cdot (c\tau + d)^k (\overline{c\tau} + d)^k v'^k \cdot du\, dv/v^2$$
$$= v^k \cdot f \cdot \bar{g} \cdot du\, dv/v^2.$$

This equation shows that the integral does not change when we integrate over $V R_G$ instead of R_G.

4. In making the following calculations it is necessary to observe that det M_j is p, not 1. If we write $M_j \tau = \tau' = u' + iv'$, $\tau = u + iv$, $M_j = (\alpha\beta|\gamma\delta)$, then $v' = pv|\gamma\tau + \delta|^{-2}$ and we easily verify that

$$v^k f(\tau)|M_j \cdot \bar{g}(\tau) = p^{-k} v'^k f(\tau')\bar{g}(\tau')|M_j^{-1}.$$

Because of the invariance of $v^{-2} du\, dv$ we now obtain the equation of the text.

Chapter III

Modular Forms with Multiplicative Fourier Coefficients. II.

In this chapter we shall continue the study of Hecke's operators and of eigenforms with multiplitative Fourier coefficients, studying them on the subgroups $\Gamma_0(n)$. A complete account of this treatment will appear in a forthcoming paper by A. O. L. Atkin and the writer*. Here we concentrate on $n = q =$ prime.

1 The Subgroup $\Gamma_0(q)$, q = Prime. Good Primes.

Let q be a fixed prime and k a fixed even positive integer. We recall that $\Gamma_0(q)$ is defined by the condition $q|c$ in $(a\, b|c\, d) \in \Gamma$. We abbreviate $\langle\Gamma_0(q), -k\rangle_0$ to $\langle\Gamma_0\rangle_0$.

The discussion is divided into two parts. First we treat the primes p that do not divide q, i.e., $p \neq q$; these are called *good*. Here the previous theory developed in the last chapter goes through almost unchanged. The operators T_p are the same as previously. The second step is to treat the *bad* prime q. The previous operator T_p is not

*This paper has since been published: Atkin, A.O.L., and J. Lehner. 1970. Hecke Operators on $\Gamma_0(m)$. *Math. Ann.* 185:134–160.

suitable, since $\langle\Gamma_0\rangle_0|T_q \not\subset \langle\Gamma_0\rangle$. We must define a new operator U_q and discuss the behaviour of U_q vis a vis $\{T_p\}$.

In the sequel p will denote any prime different from q. Let

$$H_{p,q} = H_p = \left\{ \begin{pmatrix} a & b \\ qc & d \end{pmatrix} \text{ such that } a, b, c, d \in Z, ad - qbc = p \right\}. \tag{1}$$

Then we can show that

$$H_p = \Gamma_0(q) \cdot B_p, \tag{2}$$

where B_p is the *same* set defined in Chapter II, lemma 1. Hence T_p, defined by

$$f|T_p = p^{k-1} \sum_{M \in B_p} f|M, \tag{3}$$

is identical with the operator in Chapter II, (6), (6a). Again we observe that the definition of the operator T_p is independent of the choice of representatives B_p in (2). Now exactly as before we can prove that

$$\langle\Gamma_0\rangle_0 T_p \subset \langle\Gamma_0\rangle_0,\ p \neq q. \tag{4}$$

Before proceeding further we make note of a characteristic feature of the theory. A cusp form in Γ is also a cusp form in $\Gamma_0(q)$, and since the operators $T_p (p \neq q)$ are the same for Γ and $\Gamma_0(q)$, a simultaneous eigenform for Γ is a simultaneous eigenform for $\Gamma_0(q)$. As normalization for the eigenforms in Γ_0 we shall require only that Fourier series at i^∞ should begin with $x = e^{2\pi i\tau}$, and the expansion variable x is the same for Γ and Γ_0. Thus the basis $\{h_i\}$ of normalized simultaneous eigenforms that we developed for Γ in the last chapter will serve as part of a basis for the normalized simultaneous eigenforms for Γ_0. But by Chapter II, theorem 6 the $\{h_i\}$ are unique; hence *a basis for the normalized simultaneous eigenforms for* $\Gamma_0(q)$ *must include the basis forms of* Γ.

Moreover, there is another type of imprimitivity that occurs. Let $f(\tau) \in \langle\Gamma\rangle_0$, then $g(\tau) = f(q\tau) \in \langle\Gamma_0\rangle_0$. In fact, if $(a, b|qc, d) \in \Gamma_0$,

$$g(\tau)\left|\begin{pmatrix} a & b \\ qc & d \end{pmatrix}\right. = f(q\tau)\left|\begin{pmatrix} a & b \\ qc & d \end{pmatrix}\right. = f\left|\left(q\frac{a\tau + b}{qc\tau + d}\right)\right. = f\left|\left(\frac{a(q\tau) + bq}{c(q\tau) + d}\right)\right.$$

$$= \left\{ f\left|\begin{pmatrix} a & bq \\ c & d \end{pmatrix}\right.\right\}_{q\tau} = f(q\tau) = g(\tau),$$

since $(a, bq|c, d)$ has determinant 1 and belongs to Γ. Also, writing $f(q\tau) = f(\tau)|Q$, $Q = (q\,0|0\,1)$, we observe[1] that Q commutes with $T_p (p \neq q)$. Hence

$$f(q\tau)|T_p = (f|Q)|T_p = (f|T_p)|Q = \Lambda(p) f(\tau)|Q = \Lambda(p) f(q\tau) \tag{*}$$

so that $f(q\tau)$ is a simultaneous eigenform of all T_p, $p \neq q$. *Hence the simultaneous eigenforms for* $\Gamma_0(q)$ *contain* $f(q\tau)$, *where* $f(\tau)$ *is a simultaneous eigenform* for Γ. Note however that $f(q\tau)$ is not normalized even if $f(\tau)$ is.

We shall handle the situation in the following systematic way. The scalar product of two cusp forms in Γ_0 can be defined in the same way as was done for Γ:

$$(f, g) = \int_{R_{\Gamma_0}} \int f(\tau)\bar{g}(\tau)v^{k-2} du\, dv, \tau = u + iv;$$

its existence and independence of the fundamental region R_{Γ_0} were discussed briefly in note 3 of Chapter II. Moreover, the proof that $T_p (p \neq q)$ is a hermitian operator on $\langle\Gamma_0\rangle_0$ is almost the same as before. This enables us to apply the Hecke-Petersson theory to the cusp forms on Γ_0.

But more is true. An examination of the arguments in Chapter II reveals the following: *the whole theory is valid for any subspace that is invariant under any subset of the operators T_p*. Our next move will therefore be to find subspaces of $\langle\Gamma_0\rangle_0$ that are consistent with the imprimitivities mentioned above.

Suppose M is a subspace of $\langle\Gamma_0\rangle_0$. To M there corresponds a subspace $M^\perp$, its orthogonal complement, consisting of all g in $\langle\Gamma_0\rangle_0$ for which $(g, M) = 0$ (i.e., g is orthogonal to each $f \in M$). Clearly $\langle\Gamma_0\rangle_0$ is the direct sum of M and $M^\perp$ (every form in Γ_0 is uniquely the sum of a form in M and one in $M^\perp$). Suppose M is invariant under a hermitian operator X, so that $M|X \subset M$. Then $M^\perp$ is X-invariant. Indeed, $f \in M$, $g \in M^\perp$ imply

$$(g|X, f) = (g, f|X) = 0,$$

since $f|X \subset M$; hence $g|X$ is orthogonal to all of M and thus lies in $M^\perp$. If M, N are invariant subspaces, so is their direct sum $M \oplus N$, as is obvious. Finally, if M is invariant under an operator Y, and if X commutes with Y and f is an eigenform under X, then $f|Y$ is also an eigenform with the same eigenvalue (unless $f|Y = 0$). For $(f|Y)|X = (f|X)|Y = (\lambda f)|Y = \lambda(f|Y)$. This obvious remark is often applicable.

Now we can define our subspaces. We write, as above, $f(\tau)|Q = f(q\tau)$, with $Q = (q\ 0|0\ 1)$. Define

$$\begin{aligned} A_1 &= \langle\Gamma\rangle_0 \\ A_{1q} &= \{f|Q, \text{ where } f \in A_1\}, \text{ which we write } A_1|Q \\ A &= A_1 \oplus A_{1q} \\ B &= A^\perp \text{ in } \langle\Gamma_0\rangle_0, \end{aligned} \tag{5}$$

so that

$$\langle\Gamma_0\rangle_0 = A \oplus B.$$

Lemma 1. *The subspaces A_1, A_{1q}, A, and B are each invariant under the operators T_p, $p \neq q$.*

Proof. For A_1 the assertion is known from the theory on Γ; for the other subspaces it follows from the above remarks. Q.E.D.

Lemma 1 enables us to apply the Hecke-Petersson theory separately to A_1, A_{1q}, B. We obtain in each case a basis of simultaneous eigenforms of all T_p, $p \neq q$. Write generally

$$A_1 = [h_1, \dots, h_\mu]$$

to indicate that $\{h_i\}$ is such a basis.

Lemma 2. *Let* $\mu = \dim\langle\Gamma\rangle_0$, $v = \dim\langle\Gamma_0\rangle_0$. *We have the following simultaneous-eigenform bases:*

$$\begin{aligned} A_1 &= [h_1, \dots, h_\mu] \\ A_{1q} &= [h_1|Q, \dots, h_\mu|Q] \\ B &= [f_1, \dots, f_{v-2\mu}], \end{aligned}$$

where $\{h_i\}$ *is the old basis for* $\langle\Gamma\rangle_0$.

Proof. The statement about A_1 is obvious because T_p is the same operator on Γ_0 as on Γ. Since $h \to h|Q$ is a 1–1 linear map from A_1 onto A_{1q} and Q commutes with T_p, the second statement follows. The last statement merely defines $\{f_i\}$. Q.E.D.

Lemma 3. *The set* $\{h_1, \dots, h_\mu, h_1|Q, \dots, h_\mu|Q\}$ *is linearly independent.*

Proof. If not, we have

$$-\sum_{i=1}^{\mu} c_i h_i(\tau) = \sum_{i=1}^{s} c_i' h_i(q\tau), s \leq \mu$$

for some $c_s' \neq 0$. Call the right member $\varphi(q\tau)$. Then $\varphi(q\tau) \in \langle\Gamma\rangle_0$, but also $\varphi(\tau) \in \langle\Gamma\rangle_0$. Hence $\varphi \equiv 0$, by Chapter II, lemma 3, but this contradicts the linear independence of the $\{h_i(\tau)\}$. Q.E.D.

A corollary of this lemma and (5) is that

$$[h_1, \dots, h_\mu, h_1|Q, \dots, h_\mu|Q, f_1, \dots, f_{v-2\mu}] \tag{6}$$

is a simultaneous eigenform basis for $\langle\Gamma_0\rangle_0$.

Without modifying any of the properties of the $[f_i]$ we can insist that the coefficient of the first term in the Fourier series for f_i be 1. Then we can prove that each basis form f_i is *normalized* in the sense of Chapter II, Sec. 3. We have namely,

Lemma 4.

$$f_i(\tau) = x + \dots, i = 1, \dots, v - 2\mu.$$

Proof. Suppose the lemma false. Then we can show that all terms of f_i have exponents divisible by q. Suppose not; then we write

$$f_i(\tau) = a(r)x^r + \dots + \psi_1(q\tau), r > 1,$$

where $q \nmid r$ and $\psi_1(\tau)$ is a series in x. Let $p|r$; since $p \neq q$ we can apply T_p and get

$$f_i|T_p = a(r)x^{r/p} + \dots + \psi_2(q\tau),$$

and this is a contradiction because $a(r)x^r$ was the term with smallest exponent not divisible by q. Hence all exponents in the Fourier series of f_i are multiples of q, i.e., $f_i(\tau) = \varphi(q\tau) \in \langle \Gamma_0 \rangle_0$, where $\varphi(\tau)|S = \varphi(\tau)$. Hence $\varphi(\tau) \in \langle \Gamma \rangle_0$ by ch. II, sec. 5, lemma 2, and so $\varphi \in A_1$, $f_i \in A_{1q} \subset A$. Thus $f_i \in A \cap B = 0$, a contradiction. Q.E.D.

We now introduce an equivalence relation in the set of all simultaneous eigenforms, normalized or not. Let f, g be simultaneous eigenforms. We write

$$f \underset{q}{\sim} g \tag{7}$$

to mean that f and g have the same eigenvalues with respect to all T_p except $p = q$. This equivalence relation partitions the set of all simultaneous eigenforms into disjoint equivalence classes. To each equivalence class we adjoin 0; then it becomes a subspace and so has finite dimension. Every simultaneous eigenform lies in some equivalence class and is therefore a linear combination of forms lying in that class.

Lemma 5.

$$f \underset{q}{\not\sim} g \Rightarrow (f, g) = 0.$$

Proof. Let π be such that $f|T_\pi = \lambda(\pi) f$, $g|T_\pi = \mu(\pi) g$, with $\lambda(\pi) \neq \mu(\pi)$. Then $\lambda(\pi)(f, g) = (f|T_\pi, g) = (f, g|T_\pi) = \bar{\mu}(\pi)(f, g)$. But $\mu = \bar{\mu}$, since $\bar{\mu}(g, g) = (g, g|T_\pi) = (g|T_\pi, g) = \mu(g, g)$ and, as an eigenform, $g \not\equiv 0$. Hence $\bar{\mu}(\pi) = \mu(\pi) \neq \lambda(\pi)$ and (f, g) must vanish. Q.E.D.

Lemma 6. *Let* $f(\tau) = x + \sum_{m=2}^{\infty} c(m) x^m$ *be a normalized simultaneous eigenform of all* T_p, $p \neq q$. *Then* $c(p) = \lambda(p)$, $p \neq q$.

Proof. We have, by Chapter II, theorem 2,

$$f(\tau)|T_p = \sum_{m=1}^{\infty} \left(c(pm) + p^{k-1} c\left(\frac{m}{p}\right) \right) e^{2\pi i m \tau} = \lambda(p) \sum_{1}^{\infty} c(m) e^{2\pi i m \tau},$$

the last equality coming from the fact that f is an eigenform. Hence

$$c(pm) + p^{k-1} c\left(\frac{m}{p}\right) = \lambda(p) c(m). \tag{8}$$

Since $c(1) = 1$, this implies $\lambda(p) = c(p)$ for all $p \neq q$. Q.E.D.

Note that the forms in A_{1q} are never normalized, since they begin with a term in x^q.

Lemma 7. *If* f, g *are normalized simultaneous eigenforms and* $f \underset{q}{\sim} g$, *then* f *and* g *have the same Fourier coefficients for all indices* m *not divisible by* q.

Proof. Let $f = x + \sum_{m=2}^{\infty} a(m)x^m$, $g = x + \sum_{m=2}^{\infty} b(m)x^m$. Then by lemma 6, $a(p) = \lambda(p) = b(p)$, $p \neq q$, since we are assuming $f \underset{q}{\sim} g$. Now $a(m)$ is determined uniquely, provided $q \nmid m$, from $\{a(p)\}$ by the multiplicative law (8), and the same is true for $b(m)$. Hence the conclusion. Q.E.D.

Let us now examine the equivalence classes in the subspace A defined in (5). First, h_i, $h_j (i \neq j)$ lie in different classes. Indeed, in the proof of theorem 6 of Chapter II we proved that they had distinct eigenvalues $\Lambda_i(n)$, $\Lambda_j(n)$ for at least one n with $(n, q) = 1$. Since $\Lambda_i(n)$ is determined by the $\Lambda_i(p)$ for which $p|n$ by the multiplicative relations (18), Chapter II (and likewise $\Lambda_j(n)$), it follows that there must be a prime $p(\neq q)$ such that $\Lambda_i(p) \neq \Lambda_j(p)$. Hence h_i and h_j are not equivalent.

But $h_i(\tau)$ and $h_i|Q = h_i(q\tau)$ are equivalent, for Q commutes with T_p and so

$$(h_i|Q)|T_p = (h_i|T_p)|Q = \Lambda_i(p)h_i Q.$$

Thus define the equivalence classes

$$C_i(1) = \{h_j, h_j|Q\},\ j = 1, 2, \ldots, \mu;\ Q = \begin{pmatrix} q & 0 \\ 0 & 1 \end{pmatrix}.$$

We have shown these classes are distinct. Because of lemma 3, $\dim C_j(1) \geq 2$. Finally,

$$A = C_1(1) \oplus \ldots \oplus C_\mu(1). \tag{9}$$

Next we consider the subspace B of lemma 2 from the standpoint of equivalence classes.

Lemma 8. *Let $C_j(q)$ be the equivalence class containing f_j, $j = 1, \ldots, \nu - 2\mu$. Then* $\dim C_j(q) = 1$.

Proof. We shall first show that the classes $C_j(q)$ are distinct from each other and from the classes $C_i(1)$. Suppose $f_j \underset{q}{\sim} h_i$ for some i, $1 \leqq i \leqq \mu$. Since $f_j = x + \ldots$ by lemma 4 and $h_i = x + \ldots$, it follows by lemma 7 that f_j and h_i have the same Fourier coefficients except for indices that are multiples of q. That is,

$$f_j - h_i = \varphi(q\tau)$$

and $\varphi(q\tau) \in \langle \Gamma_0 \rangle_0$, $\varphi|S = \varphi$. Hence $\varphi(\tau) \in A_1$ (ch. II, lemma 2), which implies $\varphi(q\tau) \in A_{1q}$ and so $f_j \in A$, a contradiction.

Now h_i, $h_i|Q$ are equivalent; hence $f_j \underset{q}{\not\sim} h_i|Q$, $i = 1, \ldots, \mu$.

Suppose finally $f_i \underset{q}{\sim} f_j$, $i \neq j$. Since $f_i = x + \ldots$, $f_j = x \ldots$, we have, by lemma 7,

$$f_i - f_j = \varphi(q\tau),$$

and by the usual reasoning $\varphi(q\tau) \in A$. But $f_i - f_j \in B$. Hence $f_i = f_j$.

This proves our assertion and shows that $\dim C_j(q) \geq 1$. Now

$$\nu = \dim C_1(1) + \ldots + \dim C_\mu(1) + \dim C_1(q) + \ldots + \dim C_{\nu-2\mu}(q)$$
$$= 2\mu + \nu - 2\mu = \nu.$$

Hence if any $C_j(1)$ has dimension > 2 or if any $C_j(q)$ has dimension > 1, the last equation is contradicted.

Q.E.D.

We combine (9), lemma 8, and lemma 5 to get

Theorem 1. *Let $\mu = \dim\langle\Gamma\rangle_0$, $\nu = \dim\langle\Gamma_0\rangle_0$. Let $h_1, h_2, \ldots, h_\mu$ be the normalized simultaneous eigenforms for Γ. Define the equivalence classes*

$$C_j(1) = \{h_j(\tau), h_j(\tau)|Q\},\ j = 1, 2, \ldots, \mu;\ Q = \begin{pmatrix} q & 0 \\ 0 & 1 \end{pmatrix}.$$

Then there exist $\nu - 2\mu$ additional normalized simultaneous eigenforms $f_j(\tau)$ for $\Gamma_0(q)$, such that with the equivalence classes $C_j(q) = \{f_j(\tau)\}$ we have the orthogonal decomposition

$$\langle\Gamma_0\rangle_0 = C_1(1) \oplus \ldots \oplus C_\mu(1) \oplus C_1(q) \oplus \ldots \oplus C_{\nu-2\mu}(q).$$

Furthermore,

$$\dim C_j(1) = 2,\ j = 1, 2, \ldots, \mu$$
$$\dim C_j(q) = 1,\ j = 1, 2, \ldots, \nu - 2\mu.$$

The dimensionality of $C_j(q)$ is the essential fact needed for the developments of the next section.

2 $\Gamma_0(q)$. The Bad Prime

In the present case there is only one bad prime, q itself, and it is "unramified." In place of T_q we introduce the operator U_q:

$$f|U_q = q^{k-1} \sum_{j=0}^{q-1} f| \begin{pmatrix} 1 & j \\ 0 & q \end{pmatrix} = q^{-1} \sum_{j \bmod q} f\left(\frac{\tau + j}{q}\right). \tag{10}$$

It is not hard to prove

Lemma 9. *The operator U_q commutes with all operators T_p, $p \neq q$, and*

$$\langle\Gamma_0\rangle_0 | U_q \subset \langle\Gamma_0\rangle_0.$$

We can now make good use of the last statement of theorem 1.

Theorem 2. *Let $f_j \in C_j(q)$, i.e., f_j is a "new form" on $\Gamma_0(q)$. Then f_j is an eigenform also of U_q:*

$$f_j | U_q = \lambda_j(q) f_j.$$

Proof. Since U_q commutes with $\{T_p\}$, $f_j|U_q$ is a simultaneous eigenform for all T_p with the same eigenvalue as f_j (or else is 0). Hence $f_j|U_q \underset{q}{\sim} f_j$. From $\dim C_i(q) = 1$ we conclude the result.

In other words a "new form" in $\Gamma_0(q)$ that is an eigenform for all T_p, $p \neq q$, is also an eigenform for U_q. In the proof we used only the commutativity of U_q with all T_p, $p \neq q$. Hence f_j is an eigenform of any operator W on $\langle\Gamma_0(q)\rangle_0$ that commutes with T_p, $p \neq q$. We shall make use of this remark immediately.

We can actually calculate $\lambda(q)$. Let f be a "new form" on $\Gamma_0(q)$. Define

$$W_q = \begin{pmatrix} 0 & -1 \\ q & 0 \end{pmatrix}. \tag{11}$$

It is easily verified that

$$W_q \text{ commutes with } \{T_p, p \neq q\},\ W_q^2 = -qI, \text{ and } \langle\Gamma_0\rangle_0|W_q \subset \langle\Gamma_0\rangle_0. \tag{12}$$

Hence $f|W_q = cf$, and (12) shows that

$$c = \pm(-q)^{-k/2}.$$

Use will also be made of the fact that[2]

$$f \in \langle\Gamma_0\rangle_0 \Rightarrow F = f|W_q + q^{1-k} f|U_q \in \langle\Gamma\rangle_0. \tag{13}$$

But $F_{\bar{q}} f$ and since F also is in Γ, it follows that $F \equiv 0$. Hence $c + q^{1-k}\lambda(q) = 0$,

$$\lambda(q) = \pm q^{k/2-1}. \tag{14}$$

Q.E.D.

Lemma 10.

$$\sum_{m \geqq s} a(m)x^m|U_q = \sum_{qm \geqq s} a(qm)x^m.$$

Proof omitted; it is similar to that of Chapter II, theorem 2.

If now $f(\tau) = x + \sum_{m=2}^{\infty} c(m)x^m$, we get $c(qm) = \lambda(q)c(m)$, and by (14),

$$c(q) = \lambda(q) = \gamma q^{k/2-1}, \gamma = \pm 1 \tag{15}$$

as well as

$$c(mq) = c(q)c(m), m \geqq 1. \tag{16}$$

Iteration of this equation

$$c(q^\alpha) = (c(q))^\alpha, \alpha \geqq 1. \tag{17}$$

Thus

$$\chi_q = 1 + \frac{c(q)}{q^s} + \frac{c(q)^2}{q^{2s}} \dots = 1 + \frac{c(q)}{q^s} + \left(\frac{c(q)}{q^s}\right)^2 \dots = (1 - c(q)q^{-s})^{-1}. \tag{18}$$

This[3] is the "ζ-function factor" corresponding to the "bad prime" q. Note that $c(q)$ is the same for all forms in the classes $C_j(q)$ except possibly for a $\pm$ factor. Moreover, from the multiplicative relations in $\{c(m), q \nmid m\}$ and (16), one easily deduces that the whole set $\{c(m)\}$ is multiplicative in the usual number-theoretic sense.

We summarize in a final theorem.

Theorem 3. *Let $f_j \in C_j(q)$, $j = 1, \ldots, \nu - 2\mu$, be a "new form" on $\Gamma_0(q)$ and let*

$$f_j(\tau) = x + \sum_{2}^{\infty} c_j(m) x^m.$$

Then $\{c_j(m)\}$ is multiplicative according to the laws:

$$\begin{aligned} c_j(m)c_j(n) &= c_j(mn), (m,n) = 1 \\ c_j(p^{\alpha+1}) &= c_j(p)c_j(p^{\alpha}) - p^{k-1}c_j(p^{\alpha-1}), \alpha \geqq 1, p \neq q \\ c_j(q^{\alpha}) &= (c_j(q))^{\alpha}. \end{aligned}$$

Finally

$$c_j(q) = \gamma_j q^{k/2-1}, \gamma_j = \pm 1.$$

A uniqueness theorem analogous to Th. 6 of Chapter II can also be proved: *Any form in $\langle \Gamma_0(q) \rangle_0$ that is an eigenform of all the T_p, U_q, and W_q is a constant multiple of some "new form."*

Chapter III Notes

1. Since $f(\tau + 1) = f(\tau)$ we get

$$\begin{aligned} f|T_p Q &= p^{k-1} \sum_{j=0}^{p-1} f \left| \begin{pmatrix} 1 & j \\ 0 & p \end{pmatrix} \begin{pmatrix} q & 0 \\ 0 & 1 \end{pmatrix} \right. + p^{k-1} f \left| \begin{pmatrix} p & 0 \\ 0 & 1 \end{pmatrix} \begin{pmatrix} q & 0 \\ 0 & 1 \end{pmatrix} \right. \\ &= p^{-1} \sum_{j \bmod p} f\left(\frac{q\tau + j}{p}\right) + p^{k-1} f(pq\tau) \\ &= p^{-1} \sum_{j \bmod p} f\left(\frac{q\tau + qj}{p}\right) + p^{k-1} f(pq\tau), \end{aligned}$$

since qj runs over a full residue system mod p because of $(q, p) = 1$. The last line, however, is exactly $f|QT_p$.

2. From the fact that both U_q and W_q preserve $\langle \Gamma_0(q) \rangle_0$ it follows that $F|S = F$, where $S = (1\,1|0\,1)$, since $S \in \Gamma_0$. Hence we have to prove that F is invariant

under $T = (0\,-1|1\,0)$. Now for $0 < j \leqq q-1$ we have

$$\begin{pmatrix} 1 & j \\ 0 & q \end{pmatrix}\begin{pmatrix} 0 & -1 \\ 1 & 0 \end{pmatrix} = \begin{pmatrix} j & -\dfrac{jm+1}{q} \\ q & -m \end{pmatrix}\begin{pmatrix} 1 & m \\ 0 & q \end{pmatrix} = V_m \begin{pmatrix} 1 & m \\ 0 & q \end{pmatrix},$$

where $jm \equiv -1 \pmod q$ and m may be chosen in $0 < m \leqq q-1$ and runs with j over this range. Note that $V_m \in \Gamma_0(q)$. Hence

$$\begin{aligned} q^{1-k} f \Big| U_q T &= \left\{ f \Big| \begin{pmatrix} 1 & 0 \\ 0 & q \end{pmatrix} + \sum_{j=1}^{q-1} f \Big| \begin{pmatrix} 1 & j \\ 0 & q \end{pmatrix} \right\} \Big| T \\ &= f \Big| W_q + \sum_{m=1}^{q-1} f \Big| \begin{pmatrix} 1 & m \\ 0 & q \end{pmatrix} = f \Big| W_q + q^{1-k} f \Big| U_q - f \Big| \begin{pmatrix} 1 & 0 \\ 0 & q \end{pmatrix}. \end{aligned}$$

Also

$$f \Big| W_q T = f \Big| \begin{pmatrix} -1 & 0 \\ 0 & -q \end{pmatrix} = f \Big| \begin{pmatrix} 1 & 0 \\ 0 & q \end{pmatrix},$$

since k is even. Hence $F|T = F$, as required.

3. Hecke's theory is concerned also with "ζ-functions" of a new type. Starting with a cusp form

$$f(\tau) = \sum_{m=1}^{\infty} c(m) e^{2\pi i m \tau}$$

in $\langle \Gamma, -k \rangle_0$, we associate with it the Dirichlet series

$$\varphi(s) = \sum_{m=1}^{\infty} c(m) m^{-s}.$$

Then we can prove:

(i) $\varphi(s)$ converges absolutely for $\sigma = \operatorname{Re} s > \frac{k}{2} + 1$

(ii) $\varphi(s)$ can be continued analytically to an entire function of s of finite order

(iii) if we set $R(s) = (2\pi)^{-s}\Gamma(s)\varphi(s)$, then $R(s) = (-1)^{k/2} R(k-s)$.

(We recall k is even, ≥ 2.) Conversely if φ is a Dirichlet series with the properties (i) to (iii), then $f \in \langle \Gamma, -k \rangle_0$. These assertions can be proved by introducing a "transform pair"

$$R(s) = \int_0^{\infty} f(iy) y^{s-1} dy, \quad f(iy) = \frac{1}{2\pi i} \int_{\sigma - i\infty}^{\sigma + i\infty} R(s) y^{-s} ds,$$

corresponding to the Mellin transform pair $(\Gamma(s), e^{-y})$. The functional equation (iii) is equivalent to the transformation property $f|T = f$, $T = (0\,-1|1\,0)$. For complete details cf. Hecke, Math. Ann. **112** (1936), 664–699.

(The idea was originally used by Riemann to prove the functional equation for the Riemann zeta-function. There the Dirichlet series is

$$\varphi(s) = \zeta(2s) = \sum_{m=1}^{\infty} m^{-2s} = \sum_{m=1}^{\infty} c(m)m^{-s},$$

where $c(m) = 1$ if m is a square and 0 otherwise. The associated modular form $\vartheta(\tau)$ is on a subgroup G of Γ in which the smallest translation is $\tau \to \tau + 2$:

$$\frac{1}{2}(\vartheta(\tau) - 1) = \sum_{1}^{\infty} c(m)e^{\pi i m\tau} = \sum_{1}^{\infty} e^{\pi i m^2 \tau},$$

and the additive constant 1/2 is needed because ϑ is not a cusp form. Moreover, ϑ is not fully invariant on G; rather $\vartheta | V = \epsilon,\ \vartheta,\ V \in G$, where ϵ is a root of unity. Besides this, the "degree" of ϑ is $-1/2$, a case we have not considered so far. Thus the ordinary ζ-function does not quite fit into the present scheme.)

Now suppose f is a *normalized simultaneous eigenform*. That is, $c(1) = 1$ and

$$c(mn) = c(m)\, c(n),\ (m, n) = 1 \tag{*}$$

$$c(p^{m+1}) = c(p)c(p^m) - p^{k-1}c(p^{m-1}),\ p \text{ prime}. \tag{**}$$

Because of (*) and the fact that $\sum c(m)m^{-s}$ converges absolutely we obtain

$$\varphi(s) = \sum_{1}^{\infty} c(m)m^{-s} = \prod_{\nu}\left(1 + \frac{c(p)}{p^s} + \frac{c(p^2)}{p^{2s}} + \ldots\right).$$

(Hardy-Wright, Theory of Numbers, ch. XVII, sec. 3, th. 286.) Now we wish to make use of (**). Let

$$\chi_p = 1 + \frac{c(p)}{p^s} + \frac{c(p^2)}{p^{2s}} + \ldots.$$

Remembering $c(1) = 1$ and our convention $c(p^{-1}) = 0$, we get

$$\begin{aligned}
&\chi_p \cdot \left(1 - \frac{c(p)}{p^s} + \frac{p^{k-1}}{p^{2s}}\right) \\
&= \sum_{m=0}^{\infty} \frac{c(p^m)}{p^{ms}} - \sum_{m=0}^{\infty} \frac{c(p)c(p^m)}{p^{(m+1)s}} + \sum_{m=0}^{\infty} \frac{p^{k-1}c(p^m)}{p^{(m+2)s}} \\
&= 1 + (c(p) - c(p))p^{-s} \\
&\quad + \sum_{m=1}^{\infty} \{c(p^{m+1}) - c(p)c(p^m) + p^{k-1}c(p^{m-1})\}p^{-(m+1)s} = 1.
\end{aligned}$$

Hence

$$\varphi(s) = \prod_p \chi_p = \prod_p \left(1 - \frac{c(p)}{p^s} + p^{k-1-2s}\right)^{-1}.$$

This formula is called an *Euler product*. It expresses (indeed, is equivalent to) the multiplicative laws of the $c(m)$ just as the ordinary zeta-function expresses the law of multiplication of the integers with respect to primes. Conversely, if $\varphi(s)$ is represented by an Euler product, it is easy to show that the associated modular form has multiplicative Fourier coefficients.

Since the Euler product is fully equivalent to the multiplication laws, we should get an Euler product for eigenforms on $\Gamma_0(q)$. In fact, a similar analysis reveals that for f a normalized simultaneous eigenform on $\Gamma_0(q)$ the associated Dirichlet series $\varphi(s)$ equals

$$\varphi(s) = \chi_q \prod_{p \neq q} \chi_p,$$

where χ_p is as above and $\chi_q = (1 - c(q)q^{-s})^{-1}$, as given in the text.

The result may be extended to every $\Gamma_0(m)$. Let $W = W(m) = \langle \Gamma_0(m) \rangle_0$. Suppose m' is a proper divisor of m and d a divisor of m / m'. The space $W(m')$ admits a basis of normalized simultaneous eigenforms $g_1, g_2, \ldots, g_\alpha$ with respect to all operators T_p, $(p, m) = 1$. Let A be the subspace of W spanned by the forms $g_j(d\tau)$ for all d and m', and let B be the orthogonal complement of A in W. Then B admits a basis of normalized simultaneous eigenforms, which we call *newforms*. Each newform f has a Fourier series $x + \sum_2^\infty c(m)x^m$, and $c(p) = \lambda(p)$, the eigenvalue of f under T_p. The newforms partition into equivalence classes of dimension 1, two forms being in the same class if they have the same eigenvalues for all T_p with $(p, m) = 1$. Now let q be a prime dividing m. Then $f|U_q = 0$ if $q^2|m$ and otherwise $f|U_q = -q^{k-1} f|W_q = \pm q^{k/2-1} f$. This implies

$$\varphi(s) = \prod_{q|m} \chi_q \prod_{p\nmid m} \chi_p$$

with χ_p as in the text, and $\chi_q = 1$ if $q^2|m$, otherwise $\chi_q = (1 - c(q)q^{-s})^{-1}$ with the value of $c(q)$ given in the text. An *oldform* is a form $g(d\tau)$ with g a newform on some $\Gamma_0(m')$, $m'|m$, $m' < m$, and d dividing $m|m'$. The oldforms partition into equivalence classes of higher dimension: for a given $g(\tau)$, all $g(d\tau)$ lie in one class. Forms in different classes are orthogonal. The newforms and oldforms together constitute an orthogonal basis for $\langle \Gamma_0(m) \rangle_0$. Full details and proofs appear in A. O. L. Atkin and J. Lehner, Hecke operators on $\Gamma_0(m)$, to be published in Mathematische Annalen.

This theory has an intimate connection with the theory of elliptic curves over finite fields (cf. A. Weil, Math. Ann. **168** (1967), 149–156). We say C is an elliptic curve if it can be parameterized by elliptic functions, say $x = p(u)$, $y = p'(u)$, where p is the Weierstrass function. It is known that C is nonsingular (every point on C has a tangent) and that C can be brought to the form

$$C : y^2 = x^3 + Ax + B$$

by birational transformations (those which together with their inverses are rational in the coordinates). In other words, an elliptic curve is one that is algebraically equivalent to a nonsingular cubic. Such curves are known to have genus 1.

Suppose A and B are rational integers. Let p be a prime and consider

$$C_p : y^2 = x^3 + A_p x + B_p$$

where $A \equiv A_p$, $B \equiv B_p \pmod p$. The curve C_p has coefficients in $GF(p)$. For certain primes C_p may no longer be nonsingular. Call p a *good* prime if C_p is nonsingular and a *bad* prime otherwise. It is known that for a given C there are only a finite number of bad primes. To each bad prime q we associate a positive integer e depending on the type of singularity possessed by C_q (cf. Weil, loc. cit., 155–156 for details). The integer

$$N = \prod_{q \text{ bad}} q^e$$

is called the *conductor* of C.

A point (x, y) on C_p is called a rational point if $x, y \in GF(p)$. Let N_p be the number of rational points on C_p. Obviously N_p is finite; an old result of Hasse states that $|N_p - p| < 2\sqrt{p}$. In terms of N_p, Weil defines the essential part of the zeta-function of C as follows

$$\zeta(s) = \prod_{p \text{ good}} \chi_p \prod_{q \text{ bad}} \chi_q,$$

where χ_p, χ_q are identical with those defined above provided we put $k = 2$ and

$$c(p) = p + l - N_p,\ p \text{ good}.$$

Thus we know $c(n)$ on all the primes and we calculate $c(n)$ for a general integer n by using the multiplication laws developed in the text. Set

$$f(\tau) = x + \sum_{2}^{\infty} c(n) x^n$$

with these $c(n)$. *Weil conjectures that f is a Hecke eigenform on* $\langle \Gamma_0(N), -2 \rangle_0$. There is also a reverse conjecture, from Hecke forms to elliptic curves.

There is a good deal of work being done currently on the reduction of curves mod p, a subject that has become important in diophantine analysis (cf. Cassels, J. London Math. Soc., April, 1966), and it is to be expected that the next few years will see the dissipation of some of these mysteries.

Chapter IV

Automorphisms of Compact Riemann Surfaces

1 Riemann Surfaces and Discrete Groups

The material of this chapter is not obviously related to the modular group, though, as we shall see, there is a certain connection. It also brings in concepts of analysis, but these are quickly reduced, by application of deep-lying analytic considerations, to questions of group theory. The basic reduction is due to A. Hurwitz, whose paper, Über algebraische Gebilde mit eindeutigen Transformationen in sich, Math. Ann. **41** (1893), is still widely read today.

In order to approach the subject we must widen our point of view somewhat. In chapter I we have indicated the connection of the modular group (and its subgroups) with Riemann surfaces. See particularly Note 2 to Chapter I. This connection carries over to all finitely-generated discrete groups F of real 2×2 matrices, i.e., subgroups of $\Omega = SL(2, \text{reals})$. Briefly, it is as follows. We regard F as a group of linear-fractional transformations acting on H. By identifying points in H that are equivalent under F we obtain the orbit space $\mathsf{H}/F = R$, and with the proper topology R is a Riemann surface. The identification is realized by a map $\sigma : \mathsf{H} \to R$ defined by $\tau \to F\tau$ for $\tau \in \mathsf{H}$, and we have

$$\sigma \circ f = \sigma \text{ if and only if } f \in F.$$

We can regard **H** as a covering of R, the projection map being precisely σ. If F contains elliptic elements (i.e., elements of finite order), the covering $(\mathbf{H}, \sigma)$ is "branched" at the fixed points of such elements. If F contains parabolic elements, their fixed points are real and do not lie in **H** and so their images by σ do not lie in R. They are represented, however, by "punctures" in R, each equivalence class of parabolic elements contributing one puncture. Since F is finitely generated, there are a finite number of parabolic classes and so a finite number of punctures. These punctures, if there are any, constitute the complete boundary of R. In particular, if R is compact (without punctures) or, as we say, *closed*, then any group F such that $\mathbf{H}/F = R$ cannot contain parabolic elements. If $(\mathbf{H}, \sigma)$ is an unbranched covering, F cannot contain any elliptic elements.

Now let us reverse the procedure and start out with a given *closed* Riemann surface R of genus g, where we shall assume always that

$$g \geq 2.$$

As explained on p. 18, there exists a group $F \subset \Omega$ such that $\mathbf{H}/F = R$ and such that, with the σ determined by F, the covering $(\mathbf{H}, \sigma)$ of R is unbranched. Then by what we have said, F contains no elliptic or parabolic elements. It can then be shown (Surveys, p. 236, (6)) that the presentation of F in terms of generators and relations is

$$F = \{A_1, B_1, \ldots, A_g, B_g | A_1 B_1 A_1^{-1} B_1^{-1} \ldots A_g B_g A_g^{-1} B_g^{-1} = 1\}, \tag{1}$$

where A_i, B_i are necessarily hyperbolic. F is called a *fundamental group* and it is in fact isomorphic to the fundamental group of the closed Riemann surface R of genus g.

If we now use the group F to reduce **H**, we get a Riemann surface that is at any rate conformally equivalent to R, the surface we started with.

2 Hurwitz Groups

Let us now introduce automorphisms. By an automorphism φ of R we mean a one to one conformal mapping of R on itself; φ is indeed an automorphism of the field of algebraic functions on R. The set of all automorphisms of R forms a group $G = C(R)$. This is the group we wish to study.

Let $R = \mathbf{H}/F$ and let φ be an automorphism of R. By means of the following diagram we can lift φ to a map η of **H** on itself. It can be shown that η is one to one and analytic; hence η is a real linear-fractional transformation, $\eta \in \Omega$. Here σ is the projection map mentioned above: $\sigma \circ f = \sigma$ if and only if $f \in F$. Since, by the diagram, $\varphi\sigma = \sigma\eta$, we see that $\eta f(\tau)$, and $\eta(\tau)$ are mapped by σ into the same point of R, for $\sigma\eta f = \varphi\sigma f = \varphi\sigma = \sigma\eta$. Hence $\eta f = f_1\eta$ for an $f_1 \in F$. That is, $\eta F \subset F\eta$ and similarly $F\eta \subset \eta F$. So η lies in the normalizer $N_\Omega(F) = N$ of F in Ω. The element φ of G has been lifted to an element η in the normalizer of F.

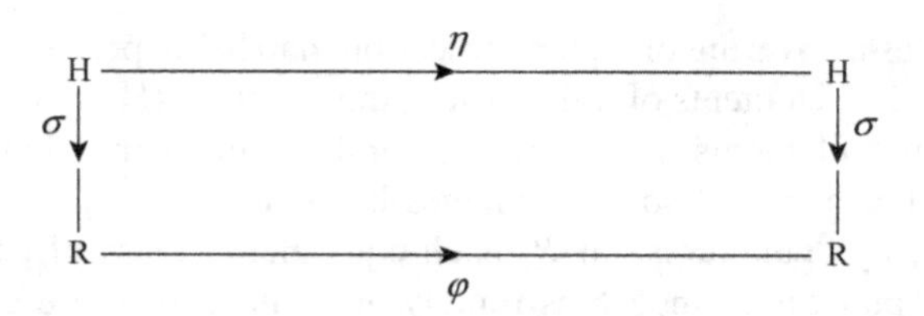

However, η is not unique. The reverse map $\eta \to \sigma\eta\sigma^{-1}$ is a homomorphism of $N \to G$ and it is onto, since every element of G can be lifted. Now $\sigma\eta\sigma^{-1} = 1$ if and only if $\sigma\eta = \sigma$, i.e., $\eta \in F$. The kernel of the homomorphism is therefore F and so

$$G \simeq N/F.$$

In other words the group G acting on R, or what is the same thing, N/F acting on $\mathbf{H}/F$, can be lifted to N acting on $\mathbf{H}$. It is therefore possible to study G by studying N/F, i.e., by studying groups of real 2×2 matrices.

It can be shown that F is not abelian, and from this it follows that N is discrete.[1] Every discrete group has a fundamental region and all fundamental regions of a given group have the same hyperbolic area, which is therefore referred to simply as the area of the group. This area can be calculated just as was done for a subgroup of Γ in Chapter I. There we presented a formula, (9), which in fact is valid for any discrete group; in particular,

$$|N| = 4\pi\left(g_N - 1 + \frac{1}{2}\left\{t_N + \sum_i \left(1 - \frac{1}{m_i}\right)\right\}\right). \tag{2}$$

Here $|N|$ is the hyperbolic area of N, t_N is the number of parabolic classes, and m_i runs over the orders of the elliptic classes in N. For example, the area of the fundamental group F is

$$|F| = 4\pi(g-1), \tag{3}$$

for we noticed in (1) that F has no parabolic or elliptic elements. Now in Chapter I, (10) we stated Siegel's result that the minimum area of a subgroup of Γ was $\pi/21$, but this result actually applies to any discrete group. (The proof consists simply in an examination of cases – cf. Short Course, p. 54.) In particular, $|N| \geqq \pi/21$. Now a fundamental region for F consists of μ copies of one for N (cf. ch. I, th. 1) and each copy has the same area since it is congruent to a fundamental region for N by a linear-fractional transformation. Here μ is the index $[N : F]$, i.e., the order of G. Thus $|F| = \mu|N|$ and we get

$$\text{order } G = \mu = |F|/|N| \leqq 4\pi(g-1)\Big/\frac{\pi}{21} = 84(g-1). \tag{4}$$

This inequality is one of Hurwitz's famous theorems. Moreover we can deduce from (2) that

$$g - 1 = \mu\left(g_N - 1 + \frac{1}{2}\left\{t_N + \sum \left(1 - \frac{1}{m_i}\right)\right\}\right); \tag{5}$$

when expressed in terms of Riemann surfaces this is called the Riemann-Hurwitz formula.

The minimum value $\pi/21$ is attained only by the group

$$\Delta = \{E_1, E_2 | E_1^2 = E_2^3 = (E_1 E_2)^7 = 1\}.$$

Formula (4) then shows that the upper bound $84(g-1)$ is attained when and only

$$N = \Delta.$$

Conversely, if F is any proper normal subgroup of Δ of finite index, then $\Delta = N_\Omega(F)$. For $N_\Omega(F)$ is certainly discrete, and if $\Delta < N_\Omega(F)$ we would have that the hyperbolic area of $N_\Omega(F)$ is smaller than that of Δ, the group with the smallest area. Moreover, F is a fundamental group. First of all F has no elements of finite order (elliptic elements). Suppose not; let F have an element E of order h, say. Then $E \in \Delta$, but we can show[2] that Δ has elements of orders 2, 3, 7 only. So suppose F has an element of order 2, say. This must be conjugate to[2] E_1. Hence E_1 itself is in F, for F is normal. That is, $\bar{E}_1$ (the image of E_1 in Δ/F) is 1. But $(\bar{E}_1\bar{E}_2)^7 = 1 = \bar{E}_2^7$, so $\bar{E}_2 = 1$ (since $\bar{E}_2^3 = 1$). This means that Δ/F is trivial or F is not proper, against our assumption. Next, F has no parabolic elements, since Δ has none. Now from the presentation of the most general finitely-generated discrete group (Surveys, p. 236, we deduce that F has the presentation (1)).

Since Δ is the normalizer of F with area $\pi/21$ and F has area $4\pi(g-1)$, it follows by previous reasoning that Δ/F has order $84(g-1)$. Hence $R = \mathbf{H}/F$ will be a surface with a "maximal automorphism group." *The problem of finding all surfaces with maximal automorphism group is therefore equivalent to finding all finite factor groups of* Δ. All this was clearly stated and proved by Hurwitz, loc. cit., and will be referred to as *Hurwitz's criterion.*

The connection with Γ arises as follows. Let $P(7)$ be the normal closure in Γ of

$$S^7 = \begin{pmatrix} 1 & 7 \\ 0 & 1 \end{pmatrix} = (TR)^7, T = \begin{pmatrix} 0 & -1 \\ 1 & 0 \end{pmatrix}, R = \begin{pmatrix} 0 & -1 \\ 1 & 1 \end{pmatrix},$$

i.e., $P(7)$ is the smallest normal subgroup of Γ that contains S^7. According to presentation theory we get a presentation of $\Gamma/P(7)$ by adding the relation $(TR)^7 = 1$ to the presentation of Γ. Thus

$$\Gamma/P(7) = \{T, R | T^2 = R^3 = (TR)^7 = 1\}, \text{ i.e., } \Gamma/P(7) = \Delta.$$

If F is a normal, finite-index subgroup of Δ we can make the following diagram:

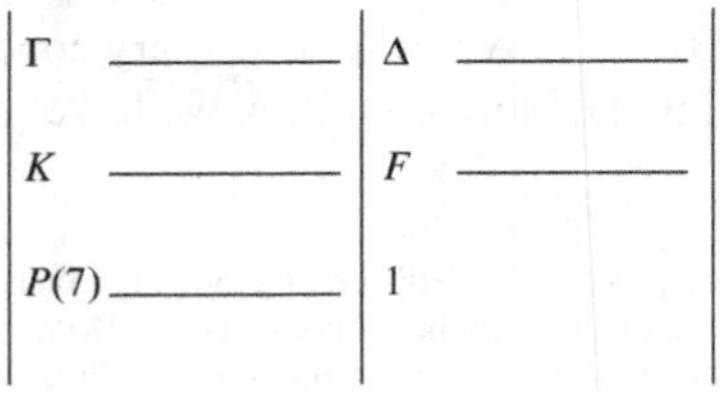

Γ ————	Δ ————	G
K ————	F ————	1
$P(7)$ ————	1	

Here K is the subgroup of Γ obtained from the inverse of the homomorphism $\Gamma \to \Delta$. We see that every maximal automorphism group G is a finite factor group of Γ by a kernel that contains S^7, and conversely. The problem of finding all surfaces with maximal automorphism group is equivalent to finding all normal subgroups of Γ that contain S^7. However, we shall approach the problem via Δ rather than Γ.

The group G is a finite factor group of Δ if and only if it is a finite group with presentation

$$G = \{x, y | x^2 = y^3 = (xy)^7 = 1, R_\alpha = 1\},$$

$R_\alpha = 1$ representing additional relations. Such a group is called a *Hurwitz group* in this chapter.

3 Linear Fractional Groups That Are Hurwitz

Our problem is now one in the theory of linear groups and one famous example was known to Klein. His simple group of order 168 has the presentation

$$\{x, y | x^2 = y^3 = (xy)^7 = (xyx^{-1}y^{-1})^4 = 1\}$$

and is therefore a factor group of Δ; it is Δ/N, where $N =$ normal closure of $(xyx^{-1}y^{-1})^4$. This group is known to be $LF(2, 7)$, the group of linear fractional transformations of determinant 1 with coefficients in $GF(7)$, the finite field of 7 elements. The general finite field, denoted by $GF(q)$, has $q = p^n$ elements, where p is a prime and n a positive integer. and we write $LF(2, q)$ for the linear group over $GF(q)$.

This raises the question: which of the $LF(2, q)$ groups are Hurwitz groups?

The group $LF(2, q)$ is simple for $p > 3$, where $q = p^n$. Many, if not all, of the known simple groups are 2-generator groups and this suggests that one should look for Hurwitz groups among the simple groups. In particular, A_n, the alternating group on n letters, is simple for $n > 4$.

It has been proved by Graham Higman that A_n is a Hurwitz group for all large n (unpublished). The remainder of this chapter is devoted to the groups $LF(2, q)$. Whether a given $LF(2, q)$ group is Hurwitz has been completely settled by A. M. Macbeath.* We shall follow Macbeath's method and determine all the $LF(2, q)$ that are Hurwitz. Here

$$q = p^n, \quad p = \text{prime}.$$

If an LF group is to be Hurwitz, it must contain elements A, B, C of orders 2, 3, 7, respectively, with $ABC = I$. A further necessary condition is that 84 divide the order of the group. Concerning the latter condition we have

---

*Fuchsian Groups, Dundee Notes, 1961. For a different proof cf. Macbeath, A.M. 1967. Generators of the linear fractional groups. Symposium on number theory. *Amer. Math. Soc.* Houston, 14–32. Cf. also Newman, M. 1968. Maximal normal subgroups of the modular group. *Proc. Amer. Math. Soc.* 19:1138–1144.

Lemma 1. *Let p^{n_0} be the smallest power of p such that 84 divides the order of $LF(2, p^{n_0})$. Then*

$$n_0 = \begin{cases} 1, & p \equiv 0, \pm 1 \pmod 7 \\ 3, & p \equiv \pm 2, \pm 3 \pmod 7 \end{cases}$$

Denote the order of $LF(2, p^n)$ by $o(p^n)$; it equals

$$o(p^n) = \delta p^n (p^{2n} - 1), \quad \delta = \begin{cases} 1, & p = 2 \\ 1/2, & p > 2 \end{cases} \tag{6}$$

Let $p > 3$. The case $p = 7$ is immediate. Suppose $p \neq 7$. If $p^2 - 1 \equiv 0 \pmod 7$ we have $o(p) \not\equiv 0 \pmod{84}$, for $p^2 - 1$ is divisible by 12 because $p > 3$. When $p \not\equiv 0, \pm 1 (7)$, $p^2 - 1$ is no longer divisible by 7, nor is $p^4 - 1$. But $p^6 - 1 \equiv 0 \pmod 7$, hence in these cases $n_0 = 3$. This proves the assertion for $p > 3$. When $p = 3$ the order is $3^n(3^{2n} - 1)/2$, which is first divisible by 7 when $n = 3$ and is also divisible by 12. Finally, for $p = 2$ the order is $2^n(2^{2n} - 1)$ and this is first divisible by 84 when $n = 3$.

The lemma shows that we need consider only the groups with $n \geq n_0$ in $q = p^n$. We shall first concentrate on $n = n_0$ and show that $LF(2, p^{n_0})$ is Hurwitz. Finally we shall prove that an $n > n_0$ does not lead to a Hurwitz group. Thus the Hurwitz groups among the $LF(2, q)$ are precisely the ones for which

$$\begin{aligned} &\text{(i)} \quad q = 7 \\ &\text{(ii)} \quad q = p \equiv \pm 1 \pmod 7 \\ &\text{(iii)} \quad q = p^3,\ p \equiv \pm 2, \pm 3 \pmod 7. \end{aligned} \tag{7}$$

We now return to the first requirement for a Hurwitz group mentioned at the beginning of this section. Let

$$q_0 = p^{n_0},\ G_0 = LF(2, q_0).$$

Lemma 2. *In G_0 there always exist elements A, B, C with*

$$A^2 = B^3 = C^7 = ABC = I. \tag{8}$$

Proof. In $LF(2, 7)$ the solution is easy:

$$A = \begin{pmatrix} 0 & 1 \\ -1 & 0 \end{pmatrix}, B = \begin{pmatrix} 0 & -1 \\ 1 & -1 \end{pmatrix}, C = \begin{pmatrix} 1 & 1 \\ 0 & 1 \end{pmatrix}.$$

Another solution is:

$$A = \begin{pmatrix} 0 & \theta \\ -1/\theta & 0 \end{pmatrix}, B = \begin{pmatrix} 0 & -\theta \\ 1/\theta & 1 \end{pmatrix}, C = \begin{pmatrix} 1 & -\theta \\ 0 & 1 \end{pmatrix},$$

where θ is a generator of the multiplicative group of $GF(7)$.

Now let $p \neq 7$. It is convenient at this point to introduce $\Omega_p = \Omega$, the algebraic closure of $GF(p)$. Since the characteristic of $GF(p)$ is not 7, there is a primitive 7th root of unity in Ω, say λ. Then $(\lambda\, 0|0\, \lambda^{-l})$ is of order 7, and so is

$$C = \begin{pmatrix} 0 & 1 \\ -1 & t \end{pmatrix}, t = \lambda + \lambda^{-1}$$

which is conjugate to the first matrix over Ω. Note further that $t^2 \neq 4$, since this would involve $\lambda = \pm 1$, but λ is a primitive 7th root.

Moreover, t lies in $GF(p^{n_0})$. To see this, consider which fields $GF(p^n)$ contain a primitive 7th root of unity. Since the multiplicative group of $GF(p^n)$ is cyclic and of order $p^n - 1$, a necessary and sufficient condition is that 7 divide $p^n - 1$. From lemma 1 we see that 7 divides $p^{2n_0} - 1$, so that $GF(p^{2n_0})$ contains a primitive 7th root of unity and therefore contains λ and λ^{-1}. Hence $t = \lambda + \lambda^{-1}$ lies in $GF(p^{n_0})$. It follows that $C \in G_0$.

Next write

$$A = \begin{pmatrix} x & y \\ z & w \end{pmatrix}, CA = \begin{pmatrix} z & w \\ -x + tz & -y + tw \end{pmatrix}.$$

We want A of order 2, CA of order 3, so

$$x + w = 0$$
$$z - y + tw = \pm 1,$$

and we have $xw - yz = 1$. Eliminating x, y we get

$$w^2 + twz + z^2 \mp z + 1 = 0.$$

This is a conic in $GF(p^{n_0})$ with discriminant $t^2 - 4 \neq 0$, and by a classical theorem[3] there are points on it. Q.E.D.

The lemma does not prove that every $LF(2, q_0)$ with q_0 satisfying (7) is a Hurwitz group. It might happen that $\{A, B\}$ – the group generated by A and B – is a proper subgroup of G_0. To determine whether $\{A, B\}$ is all of $LF(2, q)$ we shall make use of the following result, which appears in Dickson, Linear Groups:

The subgroups of $LF(2, q)$ are of the following types:

(i) cyclic and dihedral groups

(ii) tetrahedral, octahedral, and icosahedral groups

(iii) the set of elements $w = az + b$, its subgroups and their conjugates (9)

(iv) $LF(2, q')$, where $q = q'^h$, and its conjugates

(v) H, where $H/LF(2, q') = Z_2$, the group with two elements, where $q = q'^h$.

We wish to prove of course that G_0 is of type (iv) with $q' = q$. To eliminate the other possibilities we prove a series of lemmas.

Lemma 3. *A factor group other than* $\{1\}$ *of a Hurwitz group is Hurwitz.*

Let $G = \{x, y | x^2 = y^3 = (xy)^7 = 1\}$ be Hurwitz and let $G \to \bar{G}$ under a homomorphism φ. If $\varphi(x) = 1$, then $\varphi(xy) = \varphi(y)$ is of period 7, and being also of period 3, we have $\varphi(y) = 1$, so $\bar{G} = \{1\}$. By like reasoning, $\varphi(y) \neq 1$. Hence $\varphi(x)$, $\varphi(y)$ are of period 2, 3, respectively, and their product is of period 7; they therefore generate a Hurwitz group $\bar{G}$.

Lemma 4. *Let G be Hurwitz of order h. There is a simple Hurwitz group H of order dividing h.*

For let N be a maximal normal subgroup of G–there will be one unless G itself is simple. Then $G/N = H$ is Hurwitz and simple and order G = order $N \cdot$ order H.

Lemma 5. *A Hurwitz group is not solvable.*

If G (Hurwitz) were solvable, it would have an abelian factor group $H \neq \{1\}$ which, by lemma 3, is Hurwitz. But no Hurwitz group is abelian. For this would imply $(xy)^7 = x^7 y^7 = xy = 1$, so $x = y^{-1}$ and

$$x^2 = y^3 = 1 \Rightarrow x = y = 1 \text{ and } H = \{1\}.$$

Lemma 6. *The order of a Hurwitz group is divisible by 84.*

This follows from Hurwitz's criterion[4].

Lemma 7. *A Hurwitz group that is a subgroup of some* $LF(2, q)$ *is of type* (iv) *above.*

Type (i). A cyclic group is abelian and the dihedral groups are known to be solvable.

Type (ii). These groups are of order $\leqq 60$, therefore not divisible by 84.

Type (iii). Let G be the subgroup $\{w = az + b, a \neq 0\}$; G has a normal subgroup of translations H, which is cyclic. G/H is the group $\{w = az, a \neq 0\}$ and is abelian. Thus $G \supset H \supset 1$ and the factors G/H and H are abelian. So G is solvable.

Type (v). If H were Hurwitz, so would Z_2 be. Q.E.D.

The last lemma enables us to prove that G_0 is Hurwitz. There exist in G_0 elements A, B satisfying (8). The group $\{A, B\} \subset G_0$ is therefore a Hurwitz group and, by lemma 7, $\{A, B\} = LF|2, q')$, where $q' = p^n$ with $n | n_0$, in particular, $n \leqq n_0$. But by the definition of n_0 the order of $LF(2, p^n)$ is not divisible by 84 when $n < n_0$. Hence we must have $n = n_0$, i.e., $\{A, B\} = G_0$. Since $\{A, B\}$ is Hurwitz, so is G_0. We have proved

Lemma 8. *The groups*

$$G_0 = LF(2, q_0), q_0 = p^{n_0}$$

where q_0 *satisfies* (7), *are all Hurwitz groups.*

Now we must show that $LF(2, q)$, $q = p^n$ is not a Hurwitz group when $n > n_0$. For this purpose we shall partition all solutions of the matrix equation (8) into conjugacy classes.

Lemma 9. *Suppose A, B, C lie in some $LF(2, q)$ and satisfy (8). If $p \neq 7$, either the triple (A, B, C) falls into one of 3 conjugacy classes over* Ω described below or else the group $\{A, B\}$ is not Hurwitz. If $p = 7$, the triple (A, B, C) falls into one of two conjugacy classes.*

Proof. We consider first $p \neq 7$. The matrix C is diagonalizable over $\Omega (\equiv \Omega_p)$. For if not, C has equal eigenvalues. Since $\det C = 1$, the eigenvalues are α, α^{-1}, say, and equal eigenvalues mean that $\alpha = \pm 1$. Then $\pm C$ is conjugate to $(1\,1|0\,1)$ and so cannot be of order 7, since $p \neq 7$. Hence C is diagonalizable over Ω to

$$C_1 = \begin{pmatrix} \lambda & 0 \\ 0 & \lambda^{-1} \end{pmatrix},$$

λ being a primitive 7th root of unity in Ω. Let us now conjugate the relation $CA = B^{-1}$ to obtain

$$\begin{pmatrix} \lambda & 0 \\ 0 & \lambda^{-1} \end{pmatrix} \begin{pmatrix} x & y \\ z & w \end{pmatrix} = \begin{pmatrix} \lambda x & \lambda y \\ \lambda^{-1} z & \lambda^{-1} w \end{pmatrix},$$

with

$$x + w = 0$$
$$\lambda x + \lambda^{-1} w = \pm 1.$$

These equations determine x, w uniquely in Ω, since $\lambda^2 \neq 1$.

From $xw - yz = 1$ we can now determine y, z up to a further conjugacy *provided*

$$xw \neq 1.$$

For in that case $yz \neq 0$, and if

$$xw - y'z' = 1, \, y', z' \in \Omega$$

then

$$\begin{pmatrix} c & 0 \\ 0 & c^{-1} \end{pmatrix} \begin{pmatrix} x & y \\ z & w \end{pmatrix} \begin{pmatrix} c^{-1} & 0 \\ 0 & c \end{pmatrix} = \begin{pmatrix} x & y' \\ z' & w \end{pmatrix},$$

where $c = (y'/y)^{1/2} \in \Omega$. This last conjugacy preserves C_1.

However, if $yz = 0$, we have two families of solutions,

$$A = \begin{pmatrix} x & y \\ 0 & w \end{pmatrix}, \, A = \begin{pmatrix} x & 0 \\ z & w \end{pmatrix},$$

where y, z are arbitrary elements of Ω. Then the matrices A and $B = A^{-1} C_1^{-1}$ are triangular and the group $\{A, B\}$ consists of triangular matrices. By a conjugacy preserving

*$\Omega = \Omega_p$ is the algebraic closure of $GF(p)$.

C_1 we can assume the matrices are in fact superdiagonal. Thus $\{A, B\}$ is a subgroup of type (iii) in (9) and by lemma 7 is not Hurwitz.

We see then that C_1 determines A and B uniquely. Now there are 3 conjugacy classes of C_1, for C_1 is determined by the trace $\lambda+\lambda^{-1}$ and the 6 primitive roots λ yield 3 different traces.

Finally. suppose $p = 7$. Then $\pm C$ is certainly conjugate over $\Omega = \Omega_7$ to $(1\, x|0\, 1)$ and so is parabolic. It is not hard to prove[5] that every parabolic matrix in $LF(2, 7^n)$ is Ω-conjugate to one of the matrices $C_1 = \pm(1\, 1|0\, 1), \pm(1\, \theta|0\, 1)$, where θ is a generator of $LF(2, 7^n)$. With each choice of C_1 the matrices A, B are determined uniquely up to conjugacy by $A = C_1^{-1}B^{-1}$, trace $A = 0$, trace $B = \pm 1$. For

$$\begin{pmatrix} x & y \\ z & w \end{pmatrix} = \begin{pmatrix} 1 & -u \\ 0 & 1 \end{pmatrix} \begin{pmatrix} x+uz & y+uw \\ z & w \end{pmatrix}, u = 1 \text{ or } \theta$$

with

$$x + w = 0$$
$$x + w + uz = \pm 1,$$

implies $z = \mp 1/u$ and so

$$A = \begin{pmatrix} x & \pm u(x^2+1) \\ \mp 1/u & -x \end{pmatrix}.$$

Here x is arbitrary because of the conjugacy

$$\begin{pmatrix} x' & \pm u(x'^2+1) \\ \mp 1/u & -x' \end{pmatrix} = \begin{pmatrix} 1 & s \\ 0 & 1 \end{pmatrix} A \begin{pmatrix} 1 & -s \\ 0 & 1 \end{pmatrix}, s = \pm u(x - x'),$$

and this conjugacy preserves C_1. This completes the proof of the lemma. Q.E.D.

We remark that in the case $n = n_0$ there are generators of $LF(2, p^{n_0})$ in each of the conjugacy classes. In the proof of lemma 2 we can choose λ to be any primitive 7th root of unity when $p \neq 7$, and by the subsequent lemmas the A, B constructed from this λ will generate $LF(2, p^{n_0})$. Similarly, when $p = 7$ we can choose C to be $(1\, 1|0\, 1)$ or $(1\, \theta|0\, 1)$. In each case we get a triple (A, B, C) satisfying (8). Then by lemma 9 this triple lies in a unique conjugacy class, which must therefore coincide with the one determined by the original choice of λ (or of 1 or θ).

We are now ready to prove the main result.

Theorem. *An $LF(2, q)$ group is Hurwitz if and only if q satisfies* (7).

We have proved the "if" part. Conversely, it $LF(2, p^n)$ is Hurwitz, its order is divisible by 84 (by lemma 6) and this implies $n \geq n_0$ (by lemma 1). Suppose $n > n_0$; let A_n, B_n be generators of $LF(2, p^n)$. Them A_n, B_n, and $C_n = (A_n B_n)^{-1}$ satisfy the conditions (8) and, by lemma 9, the triple (A_n, B_n, C_n) falls into one of a finite number of Ω-conjugacy classes. Whatever class it may be, we can find generators A_0, B_0, C_0 of $LF(2, p^{n_0})$ that lie in the same conjugacy class. Hence $LF(2, p^n)$ is conjugate over

Ω to $LF(2, p^{n_0})$, which is clearly false since the two groups have different orders. It follows that $LF(2, p^n)$ is not Hurwitz when $n > n_0$ and the theorem is proved.

Chapter IV Notes

1. Suppose N is not discrete; let N_i be a sequence of different elements of N such that $N_i \to I$. For $A \in F$ we have $N_i A N_i^{-1} = A_i^{-1} \in F$ and $A_i^{-1} \to A$. Since F is discrete, $A = A_i'$ for all large i. Hence N_i commutes with A and so has the same fixed point set as A (Short Course, p. 9). We apply this remark to two elements A_1, A_2 of F that do not commute; such elements exist since F is not abelian. Now A_1 and A_2 have different fixed point sets (Short Course, loc. cit.). Hence for all large i, N_i must have the same fixed point set as both A_1 and A_2, which is impossible. The contradiction shows there is no sequence $N_i \to I$, and N is discrete.

2. We need the following result: every element of finite order in Δ is conjugate to an element in one of the cyclic subgroups $\{E_1\}$, $\{E_2\}$, $\{E_1E_2\}$. To prove it, we observe (Short Course. p. 55f) that there is a fundamental region R for Δ consisting of 4 sides that are paired by transformations T_1, T_2 of period 2 and 7, respectively. Defining $T_3 = T_2T_1^{-1}$, we find by examining the cycles of R that T_3 is of period 3. Each of these transformations determines a cycle; the one fixed by T_3 has two vertices; the others, a single vertex apiece. There are no other vertices in R. Suppose now $E \in \Delta$ is an element of order h; E has a fixed point α in H, and $V\alpha$ lies in $\bar{R}$ for some $V \in \Delta$. Since $V\alpha$ is a fixed point, it is a vertex of R – Short Course, p. 40 – and so it is fixed by one of T_1, T_2, T_3. Its order, then, is either 2, 3, or 7. Moreover, VEV^{-1} fixes $V\alpha$ and therefore lies in the stabilizer of $V\alpha$. In other words every element E in Δ of finite order has a conjugate in $\{T_1\}$, $\{T_2\}$, or $\{T_3\}$. In particular this is true of E_1, E_2, and E_1E_2, and the result follows.

   This result is of course only a special case. The general theorem states that in an abstract group G with presentation,

$$\{e_1, \dots, e_n; a_1, b_1, \dots, a_g, b_g | e_1^{m_1} = \dots = e_s^{m_s}$$
$$= e_1 \dots e_s a_1 b_1 a_1^{-1} b_1^{-1} \dots a_g b_g a_g^{-1} b_g^{-1} = 1\},$$

   (a so-called F-group), the "periods" $m_1, \dots, m_s$ are group invariants. That is to say, if $\bar{G}$ is a faithful representation of G as a discrete subgroup of Ω, the elements of finite order in $\bar{G}$ lie in the conjugacy classes determined by $\{E_1\}, \dots, \{E_s\}$, where $e_i \to E_i$ in the representation. Cf. Macbeath, Dundee Notes, 1961, th. 29, p. 46, and for a more general result, Lehner, Proc. Symp. on Number Theory, Amer. Math. Soc., 1965.

3. Let

$$f(x, y) = Ax^2 + Bxy + Cy^2 + Dx + Ey + F = 0$$

   be a conic with coefficients in $GF(q)$ and $\Delta = B^2 - 4AC \neq 0$. When $A = C = 0$, B cannot be 0 and f factors into

$$f/B = (x + E/B)(y + D/B) + F/B - ED/B^2,$$

and $f = 0$ has solutions in $GF(q)$. Hence we may assume $A \neq 0$ and then

$$\begin{aligned} 4Af =& (2Ax + By + D)^2 - \Delta(y + (DB - 2AE)\Delta^{-1})^2 \\ & - D^2 + 4AF + (DB - 2AE)^2\Delta^{-1}. \end{aligned}$$

The transformation

$$\begin{aligned} u &= 2Ax + By + D \\ v &= y + (DB - 2AE)\Delta^{-1} \end{aligned}$$

is invertible; hence we can solve $f = 0$ in $GF(q)$ if we can solve

$$u^2 - \Delta v^2 = G, G \in GF(q) \qquad (*)$$

in $GF(q)$.

Suppose $p > 2$. Let θ be a generator of $GF(q)$. The $(q+1)/2$ numbers u^2 for $u = 0, 1, \theta, \ldots, \theta^{(q-3)/2}$ are all distinct. This is also true of the $(q+1)/2$ numbers $\Delta v^2 + G$ for v running over the same powers of θ. However, there are only q distinct numbers in $GF(q)$ so there must be u, v such that (*) is satisfied. With a minor modification the proof works for $p = 2$ also.

4. The result that 84 divides the order of a Hurwitz group is a theorem of pure group theory, since the definition of a Hurwitz group is purely group-theoretic. It is therefore of interest to inquire whether a group-theoretic proof exists. Indeed, this is the case. The order of a Hurwitz group is immediately divisible by 42, since it contains elements of orders 2, 3, and 7. If there is a Hurwitz group whose order is not divisible by 4, there is such a group of lowest order, say H, and we can assume H is simple, because of lemma 4. The order of H is still even, for every Hurwitz group contains an element of order 2. A theorem of group theory states that the order of a simple finite group is either divisible by 12 or by the cube of the smallest prime dividing the order (in this case, 2). Cf. Marshall Hall, Jr., Theory of Groups, Macmillan, New York (1959), p. 204. Thus 4 divides the order of H.

5. Let $A = (a\,b|c\,d)$ be a parabolic matrix in $LF(2, q)$, $p > 2$. If $c = 0$ we have $A = (a\,b|0\,a^{-1}) = \pm(1\,b|0\,1)$, because of $a + a^{-1} = \pm 2$. If $c \neq 0$, A has a unique finite fixed point $x = (a-d)/2c$ and x is an element of $GF(q)$. We conjugate A to obtain, since $a + d = \pm 2$,

$$\begin{pmatrix} 0 & -1 \\ 1 & -x \end{pmatrix} \begin{pmatrix} a & b \\ c & d \end{pmatrix} \begin{pmatrix} -x & 1 \\ -1 & 0 \end{pmatrix} = \pm \begin{pmatrix} 1 & \mp c \\ 0 & 1 \end{pmatrix}.$$

If θ is a generator of $GF(q)$ we now have in both cases

$$\pm A \sim \begin{pmatrix} 1 & \theta^k \\ 0 & 1 \end{pmatrix}, 1 \leqq k < q - 1.$$

A further conjugation gives

$$\begin{pmatrix} \theta^m & 0 \\ 0 & \theta^{-m} \end{pmatrix} \begin{pmatrix} 1 & \theta^k \\ 0 & 1 \end{pmatrix} \begin{pmatrix} \theta^{-m} & 0 \\ 0 & \theta^m \end{pmatrix} = \begin{pmatrix} 1 & \theta^{k+2m} \\ 0 & 1 \end{pmatrix}.$$

By proper choice of m we make $k + 2m$ either 0 or 1 according as k is even or odd. Every parabolic matrix in $LF(2, q)$ is therefore conjugate to either $\pm(1\,1|0\,1)$ or $\pm(1\,\theta|0\,1)$, and these matrices are easily seen to be not conjugate.

Chapter V

The Fourier Coefficients of $j(\tau)$. I.

1 Introduction

So far we have considered modular forms of negative degree. There do exist forms of zero degree, i.e., modular *functions*; for example $j(\tau)$, defined in Chapter I. There are also forms of positive degree; for example, $1/\Delta(\tau)$ is of degree 12. (Cf. ch. I, sec. 3). All such forms have Fourier series expansions and it is their Fourier coefficients that we wish to discuss. Since we write $x = e^{2\pi i\tau}$, the Fourier series may be regarded as a power series in x and the Fourier coefficients are simply the coefficients $a(m)$ in the power series $\sum a(m)x^m$.

Let

$$j(\tau) = x^{-1} + c(0) + \sum_{m=1}^{\infty} c(m)x^m, x = e^{2\pi i\tau}. \tag{1}$$

Now it is known that (Surveys, pp. 349, 350)

$$c(m) \sim ae^{b\sqrt{m}}m^{-3/4}, a, b = \text{constant};$$

hence no multiplicative law such as $c(mn) = c(m)c(n)$, $(m, n) = 1$, can hold. Thus the Hecke theory cannot be extended as is to modular functions.

We have observed that $c(m)$ is an integer. A clue to the probable behaviour, arithmetically, of $c(m)$ can be had from another source. Let $p(n)$ denote the number of partitions of the positive integer n, i.e., $p(n)$ is the number of ways n can be written as a sum of positive integers, two ways being counted as the same if they differ only in the order of the summands. The famous and easily proved formula of Euler states that

$$\frac{1}{\prod_1^\infty (1-x^m)} = 1 + \sum_{n=1}^\infty p(n)x^n. \tag{2}$$

Now if we set

$$\eta(\tau) = \Delta^{1/24}(\tau) = x^{1/24} \prod_{m=1}^\infty (1-x^m), \tag{3}$$

where $x^{1/24}$ means $e^{\pi i\tau/12}$, then we can rewrite (2) as

$$x^{1/24}\eta^{-1}(\tau) = 1 + \sum_{n=1}^\infty p(n)e^{2\pi i n\tau}. \tag{4}$$

This shows us that partition problems are connected with modular forms: $p(n)$ is almost the Fourier coefficient of a "modular form of degree 1/2."

Ramanujan conjectured certain congruences for $p(n)$, in particular,

$$\begin{aligned} p(5n+4) &\equiv 0 \pmod 5 \\ p(7n+5) &\equiv 0 \pmod 7),\ n = 0, 1, 2, \ldots. \end{aligned} \tag{5}$$

He also gave proofs of (5) and so did many other authors. The feature of these congruences is that the arguments form an arithmetic progression of difference equal to the modulus of the congruence.

Ramanujan also suggested an analytic method of proving these congruences, namely, by establishing certain *identities*, of which the following is a sample:

$$\sum_{n=0}^\infty p(5n+4)x^n = 5\,\frac{\prod(1-x^{5m})^5}{\prod(1-x^m)^6}.$$

An immediate consequence of this equation is the first congruence of (5). When the equation is rewritten in terms of the η-function by means of (4), it becomes

$$\sum_{l=0}^4 \eta(5\tau)\eta^{-1}\left(\frac{\tau+24l}{5}\right) = 5^2\left(\frac{\eta(5\tau)}{\eta(\tau)}\right)^5. \tag{6}$$

This is an identity between modular *functions* belonging to a certain subgroup of the modular group. Equation (6) is proved in Rademacher, The Ramanujan identities under modular substitutions, Trans. Amer. Math. Soc. **51** (1942).

Consideration of the above problem is complicated by the fact that η is a form of fractional degree – a concept that can be accurately defined – and this implies that η has "multipliers":

$$\eta | V = \epsilon \eta, \ V \in \Gamma$$

where $\epsilon = \epsilon(V)$ is a certain very complicated 24th root of unity. From the standpoint of function theory it is much simpler to treat the simplest possible modular form, namely $j(\tau)$, which is of zero degree and has no multipliers. This is what we shall do, and our first object will be to look for congruences for the Fourier coefficients of $j(\tau)$.

2 Basic Lemmas

The modular properties of j are not affected by the addition of a constant; we shall therefore rewrite

$$j(\tau) = x^{-1} + \sum_{m=1}^{\infty} c(m) x^m, \ x = e^{2\pi i \tau} \tag{7}$$

as a normalization. We shall make use of the operator U_q defined in formula (10) of Chapter III, sec. 2; according to lemma 10 of that section U_q has the following effect:

$$j | U_q = \sum_{m=1}^{\infty} c(qm) x^m, \tag{8}$$

since we have set $c(0) = 0$ in (7). Hence we can study $c(qm)$ – the coefficients of an arithmetic progression of indices – by studying $j | U_q$.

We shall see later that $j | U_q$ belongs to the subgroup $\Gamma_0(q)$. The theory is very much simplified if $\Gamma_0(q)$ is of genus zero. In this connection we have

Lemma 1. *When $q =$ prime, $\Gamma_0(q)$ is of genus 0 if and only if*

$$q = 2, 3, 5, 7, 13.$$

We shall prove only the "if" part[1], but before doing so, let us make a few remarks about $\Gamma_0(q)$. It is easier to treat $\Gamma^0(q)$ and just as effective, since $\Gamma_0(q) = T \Gamma^0(q) T^{-1}$, $T = (0 \, -1 | 1 \, 0)$. We verify that

$$\Gamma = \Gamma^0(q) \cdot B,$$

where

$$B = \left\{ \begin{pmatrix} 1 & j \\ 0 & 1 \end{pmatrix}, 0 \leqq j < q; \begin{pmatrix} 0 & -1 \\ 1 & 0 \end{pmatrix} \right\},$$

so

$$[\Gamma : \Gamma_0(q)] = q + 1.$$

Hence the fundamental region $R(\Gamma^0) = \cup_{b \in B} b R_\Gamma$ has two inequivalent parabolic cusps, namely, ∞ and 0. Let e_2, e_3 be the number of elliptic cycles of orders 2, 3 in $R(\Gamma^0)$. Since each region bR_Γ has the same area (ch. I, (9)), we deduce from the hyperbolic area formula

$$g-1+\frac{1}{2}\left(2+\frac{1}{2}e_2+\frac{2}{3}e_3\right)=\frac{q+1}{12},\quad g+\frac{e_2}{4}+\frac{e_3}{3}=\frac{q+1}{12},\tag{*}$$

where g = genus of $\Gamma^0(q)$. Hence $q \leqq 7$ implies $g < 1$, i.e., $g = 0$. Suppose $q = 13$. Then $g \leqq 1$, but $g = 1$ implies $3e_2 + 4e_3 = 2$, an impossible equation in nonnegative integers e_2, e_3. The same results clearly hold for $\Gamma_0(q)$. Q.E.D.

Let $\Gamma_0(q)$ be of genus 0. Then there exists a Hauptmodul* $\Phi_q(\tau) = \Phi$, which we can determine completely by demanding that it have a simple zero at $\tau = i^\infty$ with residue 1 and a simple pole at $\tau = 0$. The normalized Φ will have integral coefficients, as we shall see. Because $j|U_q \in \{\Gamma_0(q), 0\}$, it can be expressed as a rational function of Φ, and from this identity we can deduce our congruence.

Lemma 2.

$$\{\Gamma_0(q), 0\}|U_q \subset \{\Gamma_0(q), 0\}.$$

Proof. Let $f \in \{\Gamma_0(q), 0\}$. Then for $V = \begin{pmatrix} a & b \\ qc & d \end{pmatrix} \in \Gamma_0(q)$,

$$(f|U_q)|V = q^{-1}\sum_{l=0}^{q-1} f\left|\begin{pmatrix} 1 & l \\ 0 & q \end{pmatrix}\begin{pmatrix} a & b \\ qc & d \end{pmatrix}\right..$$

Now

$$\begin{pmatrix} 1 & l \\ 0 & q \end{pmatrix}\begin{pmatrix} a & b \\ qc & d \end{pmatrix} = \begin{pmatrix} a+lqc & b+ld \\ q^2c & qd \end{pmatrix} = \begin{pmatrix} a+lqc & \dfrac{b+ld-am}{q}-lcm \\ q^2c & d-qcm \end{pmatrix}\begin{pmatrix} 1 & m \\ 0 & q \end{pmatrix},$$

where m is determined by $am \equiv b+ld \pmod q$. This congruence is solvable, as $(a,q) = 1$. Since also $(d,q) = 1$, we can solve for l, given m; hence m runs with l over $0, 1, \ldots, q-1$. Therefore

$$(f|U_q)|V = q^{-1}\sum_{m=0}^{q-1} f\left|\begin{pmatrix} 1 & m \\ 0 & q \end{pmatrix}\right. = f|U_q.$$

Obviously $f|U_q$ is holomorphic in $\mathbf{H}$, since f is. Thus conditions i) and iii) of the definition of a modular function (cf. (19), ch. I) are satisfied. In order to show that condition (ii) is satisfied, we shall develop expansions for $f|U_q$ at i^∞ and 0, which we need in any case. Q.E.D.

*A Hauptmodul on a group G is a univalent function on G (cf. ch. I, th. 4).

Let $\varphi(\tau) = f|U_q$. Suppose we have proved the existence of expansions for φ at $\tau = i^\infty$ and $\tau = 0$, i.e., we have

$$\varphi|V_1 = \sum_{m\geq n} \alpha(m)e^{2\pi i m\tau/\lambda}, \quad \varphi|V_2 = \sum_{m\geq n} \beta(m)e^{2\pi i m\tau/\mu}$$

where $V_1, V_2 \in \Gamma$, $V_1\infty = \infty$, $V_2\infty = 0$, and λ, μ are positive integers. For example we could choose $V_1 = I$, $V_2 = T = (0 - 1|1\,0)$. (If we had selected a $W \in \Gamma$ instead of V_1 with $W\infty = \infty$, the expansion of $f|W$ would have different coefficients $\alpha'(m)$ but λ and μ would be unchanged.) Suppose now A is an integral matrix of determinant $n > 0$, as specified in (19. ii). Then $A\infty$ is a cusp and so is equivalent by $\Gamma_0(q)$ to i^∞ or 0; say $W_iA\infty$ is i^∞ or 0 as $i = 1$ or 2, $W_i \in \Gamma_0(q)$. If we set $A_i = W_iA$, we have that $V_i^{-1}A_i$ fixes ∞ and is integral of determinant n, so that

$$D_i = V_l^{-1}A_i = \begin{pmatrix} r & t \\ 0 & s \end{pmatrix}, rs = n.$$

Since $\varphi|A = \varphi|W_i^{-1}A_i = \varphi|A_i$, we get

$$\varphi|A = \varphi|V_iD_i = (\varphi|V_i)\left(\frac{r\tau + t}{s}\right),$$

where we note $r/s > 0$. It follows that

$$\lim_{\tau\to i^\infty} \varphi(\tau)|A = \lim_{\tau\to i^\infty} \varphi|V_i(\tau)$$

exists. Thus in order to satisfy requirement (ii) of the definition, it is sufficient to exhibit expansions for $\varphi = f|U_q$ at i^∞ and 0.

At i^∞ we have the expansion $f(\tau) = \sum_{m\geq s} a(m)x^m$, hence by lemma 10, Chapter III,

$$f|U_q = \sum_{qm\geq s} a(qm)x^m, \tag{9}$$

an expansion of the required form.

Next consider the cusp $\tau = 0$. We shall need

Lemma 3.

$$qf|U_q - qf|U_qT = f\left(\frac{\tau}{q}\right) - f\left(-\frac{1}{q\tau}\right), T = \begin{pmatrix} 0 & -1 \\ 1 & 0 \end{pmatrix}.$$

Proof. The assertion is equivalent to

$$\sum_{1}^{q-1} f(M_{l\tau}) = \sum_{1}^{q-1} f(M_lT_\tau), M_l = \begin{pmatrix} 1 & l \\ 0 & q \end{pmatrix}.$$

But

$$M_l T = \begin{pmatrix} 1 & l \\ 0 & q \end{pmatrix}\begin{pmatrix} 1 & -1 \\ 1 & 0 \end{pmatrix} = \begin{pmatrix} l & -1 \\ q & 0 \end{pmatrix}$$
$$= \begin{pmatrix} l & -\frac{lm+1}{q} \\ q & -m \end{pmatrix}\begin{pmatrix} 1 & m \\ 0 & q \end{pmatrix}, lm \equiv -1 \pmod q,$$

so

$$\sum_1^{q-1} f(M_l T_\tau) = \sum f\left(\begin{pmatrix} l & \cdot \\ q & -m \end{pmatrix}\begin{pmatrix} 1 & m \\ 0 & q \end{pmatrix}\tau\right)$$
$$= \sum_1^{q-1} f(M_m \tau),$$

as required. Q.E.D.

Continuing with the proof of lemma 2, we recall that f is on $\Gamma_0(q)$ and so there is an expansion at $\tau = 0$, which according to the discussion near Chapter I (24b) is

$$f|T = f(-1/\tau) = \sum_{m \geq t} b(m) e^{2\pi i m\tau/q},$$

since $TS^q T = (1\,0|-q\,1) \in \Gamma_0(q)$. Replacing τ by $q\tau$, we get,

$$f|W = f(-1/q\tau) = \sum_{m \geqq t} b(m) x^m, x = e^{2\pi i\tau}, W\tau = -1/q\tau$$

and this is the expansion we shall use in future at $\tau = 0$. (In other words, we are using $\exp(-2\pi i|q\tau)$ as a local variable at $\tau = 0$.) Hence, using lemma 3 with τ replaced by $q\tau$, we get

$$(-qf|U_q)_{-1/q\tau} = (-qf|U_q)|W$$
$$= -q \sum_{qm \geqq s} a(qm) x^{qm} + \sum_{m \geqq s} a(m) x^m - \sum_{m \geqq t} b(m) x^{qm}. \qquad (10)$$

Equations (9), (10) show that $f|U_q$ has the behavior required of a modular form at the cusps 0, i^∞ of the fundamental region of $\Gamma_0(q)$, and lemma 2 is proved. In particular $j(\tau)|U_q$ is on $\Gamma_0(q)$.

Lemma 4. *We have*

$$\Phi|W = q^{-r/2}\Phi^{-1}, W\tau = -1/q\tau. \qquad (11)$$

The proof is immediate once we know the formula

$$\eta(\tau)|T = (-i)^{1/2}\eta(\tau), \quad (-i)^{1/2} = e^{-\pi i/4}. \qquad (12)$$

For the proof of (12) cf. C. L. Siegel, Mathematika, 1954; a sketch appears in [3].

Lemma 5. *Let $r(q-1)=24$, $q=2,3,5,7,13$. Then*

$$\Phi_q(\tau)=\Phi=\left(\frac{\eta(q\tau)}{\eta(\tau)}\right)^r=x+d_2x^2+\dots \tag{13}$$

is a Hauptmodul on $\Gamma_0(q)$, Φ has a simple zero at i^∞, a simple pole at 0, and its Fourier series has integral coefficients.

All statements follow easily from (3) and lemma 4 except the invariance on $\Gamma_0(q)$. That is proved in [2].

3 Congruences for $c(m)$ mod q

The next step is to express the modular functions on $\Gamma_0(q)$ as polynomials in Φ and Φ^{-1}.

Lemma 6. *Let $f\in\{\Gamma_0(q),0\}$ and let*

$$f(\tau)=\sum_{m=8}^{\infty}a(m)x^m,\ f|W=f(-1/q\tau)=\sum_{m=t}^{\infty}b(m)x^m$$

with integral $a(m)$, $b(m)$. Then

$$f(\tau)=\text{const.}+\sum_{l\geq 1}C_l q^{\frac{rl}{2}}\Phi^l+\sum_{h\geq 1}C_{-h}\Phi^{-h} \tag{14}$$

with integral C_l, C_{-h}, the sums being finite.

Define

$$\delta(\tau)=f-\sum_{l\geq 1}C_l q^{\frac{rl}{2}}\Phi^l-\sum_{h\geq 1}C_{-h}\Phi^{-h},$$

with undetermined C_l, C_{-h}. The expansions of Φ, Φ^{-1} at the cusps 0, i^∞ are known from (11) and (13):

$$\Phi(\tau)=x+O(x^2),\ \Phi^{-1}(\tau)=x^{-1}+O(1),$$
$$q^{\frac{r}{2}}\Phi|W=x^{-1}+O(1),\ \Phi^{-1}|W=O(x),$$

all series having integral coefficients. Since trivially $Q^j|W=(\Phi|W)^j$, these equations show that any negative power x^{-j} with an integral coefficient may be removed by subtracting a suitable integral multiple of $q^{\frac{rj}{2}}\Phi^j|W$ or of Φ^{-j} (the two being equal).

Now

$$\delta|W = f|W - \sum_{l\geq 1} C_l\Phi^{-l} - \sum_{h\geq 1} C_{-h} q^{\frac{rh}{2}} \Phi^h.$$

Since $\Phi^h = x^h + \ldots$, we can find integers $C_l, l \geq 1$ so that

$$\delta|W = b(0) + O(x).$$

Then we can find integers $C_{-h}, h \geq 1$, so that

$$\delta = a(0) + O(x).$$

Hence $\delta(\tau)$ is a modular function that is regular in **H** and at the cusps of the fundamental region and so is a constant (ch. I, th. 3). Q.E.D.

Observe that if $s \geq 0$, all C_{-h} are zero, while if $t \geq 0$, all C_l are zero. At least one of s, t must be negative if f is not to be a constant.

It is now a simple matter to apply the lemma to $qj|U_q$. The expansion of this function at i^∞ is given by (8), so that in the lemma ($f = qj|U_q$) we have $s = 1$. Also $j(\tau)$ is invariant under $T\tau = -1/\tau$, so

$$j(\tau)|W = j(-1/q\tau) = j(q\tau) = x^{-q} + c(1)x^q + \ldots;$$

hence, from (10), with $f = j$,

$$(qj|U_q)|W = x^{-q^2} - x^{-1} + O(x),$$

with integral coefficients. Thus $t = -q^2$. By the remark following the lemma, all C_{-h} are zero. It follows that the constant in (14) is zero, since Φ begins with x. Hence, dividing by q, we obtain

$$j|U_q = \sum_{m=1}^{\infty} c(qm)x^m = \sum_{l=1}^{q^2} C_l q^{\frac{rl}{2}-1}\Phi^l, \tag{15}$$

and Φ^l has integral coefficients. Thus $j|U_q$ is equal to $q^{\frac{r}{2}-1}$ times a power series with integral coefficients. Using the values of r in the five cases, we obtain the following *congruences*:

$$\begin{aligned} c(2m) &\equiv 0 \pmod{2^{11}} \\ c(3m) &\equiv 0 \pmod{3^5} \\ c(5m) &\equiv 0 \pmod{5^2} \\ c(7m) &\equiv 0 \pmod{7},\ m = 1, 2, \ldots \end{aligned} \tag{16}$$

Since $r = 2$ for $q = 13$, there is no congruence for $q = 13$; we shall discuss this situation more completely in the next chapter.

4 Congruences for $c(m)$ mod q^α

We shall regard (16) as the first step of an induction that will result in congruences to powers of q. First, however, we must discuss the *modular equation*.

Theorem 1. *Let $G \subset \Gamma$, $[\Gamma : G] < \infty$, and suppose f, g are two modular functions on G. Then there is a polynomial P with complex coefficients such that, identically in τ,*

$$P(f(\tau), g(\tau)) = 0.$$

If we recall that f, g project into meromorphic functions χ, ψ on the Riemann surface $\mathbf{H}/G$, the theorem becomes evident, for $\mathbf{H}/G$ becomes a compact surface when the punctures arising from the parabolic classes of G are filled in. In the neighbourhood of a puncture p, $\chi(q)$ is meromorphic and χ tends to a limit (finite or infinite) as $q \to p$ because of the similar condition placed on $f(\tau)$ as τ tends to a parabolic cusp. But the set of meromorphic functions on a compact Riemann surface forms an algebraic function field in one variable, and any two elements of the field are connected by an algebraic equation, which induces the same algebraic equation satisfied by f, g. A direct proof that remains in $\mathbf{H}$ can be found in Short Course, p. 90.

The algebraic equation connecting f, g is called a *modular equation*. Set

$$Y(\tau) = q^{\frac{r}{2}} \Phi_q \left(\frac{\tau}{q}\right). \tag{17}$$

Then $Y \in \{\Gamma^0(q), 0\}$, for with $V = (a, qb|c, d) \in \Gamma^0(q)$,

$$\Phi\left(\frac{V\tau}{q}\right) = \Phi\left(\frac{a(\tau/q)+b}{qx(\tau/q)+d}\right) = \Phi\left(\frac{\tau}{q}\right)$$

since Φ is invariant on $\Gamma_0(q)$. Now $\Gamma_0(q)$, $\Gamma^0(q)$ have the common subgroup

$$\Gamma_0^0(q) = \left\{\begin{pmatrix} a & b \\ c & d \end{pmatrix} \in \Gamma \text{ such that } b \equiv c \equiv 0 \pmod q\right\}.$$

of finite index in both. Hence Φ and Y are connected by a modular equation.

To find the modular equation use lemma 6 to write

$$j(\tau) = \Phi^{-1}(\tau) + \sum_{l=0}^{q} a_l \Phi^l(\tau),\ \Phi = \Phi_q, a_l = a_l(q). \tag{18}$$

Now the left member of (18) is invariant under $\tau \to -1/\tau$, so the right hand side must be, and when we apply (11) we get

$$Y + \sum_{l=1}^{q} a_l Y^{-1} = \Phi^{-1} + \sum_{l=1}^{q} a_l \Phi^l,$$

$$Y - \Phi^{-1} + \sum_{l=1}^{q} a_l (Y^{-1} - \Phi^j) = 0.$$

The left member has the factor $Y^{-1} - \Phi$; excluding this and multiplying by Y^{q-1} we obtain, after rearrangement,

$$Y^q + \sum_{l=1}^{q} (-1)^l p_l Y^{q-1} = 0. \tag{19}$$

with

$$(-1)^{l+1} p_l = q^{r/2+2} \sum_{k=l}^{q} b_k \Phi_q^{k-l+1}, \tag{20}$$

where we have written $a_k = q^{r/2+2} b_k$. The b_k can be calculated and it turns out that except in one case $b_k = b_k(q)$ is an integer.

We now consider the typical prime $q = 5$ and find by actual calculation:

k	1	2	3	4	5
b_k	63	52.5^3	63.5^5	6.5^8	5^{10}

(21)

The original work on this problem appears in Lehner, *Amer. J. of Math.* (1949), 136–148; 373–386. Improvements in technique, as well as important new results, were introduced in Atkin and O'Brien, *Trans. Amer. Math. Soc.* (1967). We shall follow the methods of the latter paper.

The remainder of this chapter, as well as all of the next, depends on the systematic use of a certain valuation.

For a positive integer a, let $\pi(a)$ be the largest power of 5 that divides a, i.e., $\pi(a) = e$ if and only if $5^e \nmid a$, $5^{e+1} \nmid a$. We define $\pi(a/b) = \pi(a) - \pi(b)$; finally set $\pi(0) = \infty$. Thus $\pi(0) \geqq m$ for all positive integers m. Clearly

$$\pi(ab) = \pi(a) + \pi(b),$$
$$\pi(a+b) \geqq \min(\pi(a), \pi(b))$$

with equality if $\pi(a) \neq \pi(b)$. For example, (21) gives $\pi(b_1) = 0, \pi(b_2) = 3, \ldots, \pi(b_5) = 10$.

The general idea of the method we are going to use is as follows. By iterating (15) we obtain*

$$j|U_5^\alpha = \sum_{n=1}^{\infty} c(5^\alpha n)x^n = \sum_{l \geq 1} C_l 5^{3l-1} \Phi^l | U_5^{\alpha-1}, \quad \alpha \geq 1 \tag{21a}$$

where $U_5^\alpha = U_5^{\alpha-1}|U_5$. The inductive step from α to $\alpha+1$ is made by applying U_5, and obviously we must treat $\Phi^l|U_5$. Since $\Phi(\tau+1) = \Phi(\tau)$, we see from (19) and (20) that

*In what follows we often omit the upper limit of a sum, the understanding being that all sums are finite.

the roots of (19) are $Y(\tau+\nu)$, i.e.,

$$5^3\Phi_5\left(\frac{\tau+\nu}{5}\right), \nu = 0, \ldots, 4. \tag{*}$$

Hence

$$\Phi|U_5 = 5^{-1}\sum_{\nu=0}^{4}\Phi\left(\frac{\tau+\nu}{5}\right) = 5^{-4}S_1 = 5^{-4}p_1,$$

where, in general,

$$S_h = \sum_{\nu=0}^{4}\left(5^3\Phi\left(\frac{\tau+\nu}{5}\right)\right)^h \tag{22}$$

is the sum of the hth powers of the roots. To calculate S_h we shall use Newton's formula for the sums of powers of the roots of an algebraic equation.

First,

$$S_1 = p_1 = 5^3\sum_{k=1}^{5} b_k\Phi^k = \sum_{k=1}^{5} c_{1k}\Phi^k,$$

this being a definition of c_{1k}. Since $c_{1k} = 5^5 b_k$, we verify from (21) that

$$\pi(c_{1k}) = 5 + \pi(b_k) \geqq \left[\frac{5(k+1)+1}{2}\right]. \tag{23}$$

This is the case $h = 1$ of lemma 7, which we are about to state.

Let c_{hk} be defined by

$$S_h = \sum_{k\geqq 1} c_{hk}\Phi^k. \tag{24}$$

That S_h can be written in this form follows by induction from Newton's formula (25) below.

Lemma 7.

$$\pi(c_{hk}) \geqq \left[\frac{5(k+1)+1}{2}\right].$$

This is the crucial lemma needed to establish the inductive step from α to $\alpha+1$.

To prove the lemma we recall that Newton's formula is

$$S_h = \sum_{j=1}^{h}(-1)^{j+1} p_j S_{h-j}, h \geqq 1 \tag{25}$$

with $p_j = 0$ for $j > 5$, $S_0 = h$. Now assume the lemma for c_{ik} when $i < h$ and all k. Then from (20), (24), (25),

$$\begin{aligned} S_h &= \sum_{t\geqq 1} c_{ht}\Phi^t = \sum_{j=1}^{h}(-1)^{j+1} p_j S_{h-j} \\ &= \sum_{j=1}^{h-1} 5^5 \sum_{i=j}^{5} b_i \Phi^{l-j+1} \sum_{k\geqq 1} c_{h-j,k}\Phi^k + h\cdot 5^5 \sum_{i=h}^{5} b_i \Phi^{l-h+1}. \end{aligned}$$

so that

$$c_{ht} = 5^5 h b_{t+h-1} + 5^5 \sum_{j\geqq 1}\sum_{k\geqq 1} c_{h-j,k} b_{t-k-1+j}. \tag{26}$$

Hence, using $\pi(b_j) \geqq \left[\frac{5j-4}{2}\right]$ (as verified from (21)) and the properties of the valuation π, we obtain

$$\pi(c_{ht}) \geqq \min\left(5 + \left[\frac{5(t+h-1)-4}{2}\right], 5 + \psi(h,t)\right),$$

where

$$\psi_{h,t} \geqq \min_{j,k}\left(\left[\frac{5(k+h-j)+1}{2}\right]\left[\frac{5(t-k+j-1)-4}{2}\right]\right).$$

We now use a characteristic argument. Write the expression in the right member as

$$\min_{k-j=\lambda}\left(\left[\frac{5(\lambda+h)+1}{2}\right]\left[\frac{5(t-\lambda)-9}{2}\right]\right);$$

it is invariant under $\lambda \to \lambda + 2$. Now in the right member of (26) the index $\nu = t - k - 1 + j$ of b_ν must be $\geqq 1$ so $\lambda \leqq t - 2$. Hence the minimum is attained for $\lambda = t - 3$ or $\lambda = t - 2$:

$$(\quad)_{\lambda=t-3} = \left[\frac{5(t+h)-14}{2}\right] + 3, (\quad)_{\lambda=t-2} = \left[\frac{5(t+h)-9}{2}\right]$$

and so for $\lambda = t - 2$. Therefore $\psi_{h,t} \geqq \left[\frac{5(t+h)-9}{2}\right]$, hence

$$\pi(c_{ht}) \geqq \min\left(\left[\frac{5(t+h)+1}{2}\right], \left[\frac{5(t+h)+1}{2}\right]\right) = \left[\frac{5(t+h)+1}{2}\right],$$

completing the induction and the proof of the lemma.

Since by (22)

$$S_h = 5^{3h+1}\Phi^h|U_5,$$

we get, as an immediate corollary:

$$\pi(a_{hk}) \geqq \left[\frac{5k-h-1}{2}\right], \tag{27}$$

where we have set

$$\Phi^h|U_5 = \sum_{k\geqq 1} a_{hk}\Phi^k. \tag{28}$$

Thus $\Phi^h|U_5$ is a polynomial in Φ and by induction the same is true of $\Phi^h|U_5^\alpha$.

If we apply the last statement to (21a) we see that $j|U_5^\alpha$ is a polynomial in Φ and we write

$$j(\tau)|U_5^\alpha = \sum_1^\infty c(5^\alpha n)x^n = \sum_{r\geqq 1} j_{\alpha r}\Phi^r. \tag{29}$$

The congruence we wish to prove would follow from $\tau(j_{\alpha r}) \geqq \alpha+1$. In order to prove this, however, we have to make the stronger assumption

$$\pi(j_{\alpha r}) \geqq \alpha + 1 + \left[\frac{5r-5}{2}\right] \text{ for an } \alpha \text{ and all } r \geqq 1. \tag{30}$$

For $\alpha = 1$, (30) is verified from (15):

$$\pi(j_{1r}) = \pi(C_r 5^{3r-1}) \geqq 3r-1 \geqq 2 + \left[\frac{5r-5}{2}\right].$$

Apply U_5 to (29):

$$j(\tau)|U_5^{\alpha+1} = \sum_{r\geqq 1} j_{\alpha r}\Phi^r|U_5 = \sum_{r\geqq 1} j_{\alpha r}\sum_{k\geqq 1} a_{rk}\Phi^k,$$

implying

$$j_{\alpha+1,r} = \sum_{k\geqq 1} j_{\alpha k}a_{kr}.$$

Hence, using (30) and (27)

$$\pi(j_{\alpha+1,r}) \geqq \alpha + 1 + \min_k\left(\left[\frac{5r-k-1}{2}\right] + \left[\frac{5k-5}{2}\right]\right).$$

The expression in parentheses is increasing for $k \to k+2$ and its minimum occurs for $k = 1$; hence

$$\pi(j_{\alpha+1,r}) \geqq \alpha + 1 + \left[\frac{5r-2}{2}\right] \geqq \alpha + 2 + \left[\frac{5r-5}{2}\right].$$

This completes the proof of

Theorem 2.

$$c(5^\alpha n) \equiv 0 \pmod{5^{\alpha+1}}, \alpha \geqq 1, n \geqq 1.$$

The results for the remaining primes are:

$$\begin{aligned} c(2^\alpha n) &\equiv 0 \pmod{2^{3\alpha+8}} \\ c(3^\alpha n) &\equiv 0 \pmod{3^{2\alpha+3}} \\ c(7^\alpha n) &\equiv 0 \pmod{7^\alpha}, \alpha \geqq 1, n \geqq 1 \end{aligned} \tag{31}$$

Chapter V Notes

1. To prove that the primes mentioned are the only ones for which Γ_0 is of genus zero, we shall develop a formula for the genus. As in the text we treat Γ^0 instead of Γ_0.

   Let α be an elliptic fixed point of Γ^0 of order 2 and let G_α be its stabilizer; G_α is cyclic of order 2. If β is another fixed point of order 2, then β and α are Γ^0-equivalent if and only if G_β, G_α are conjugate by Γ^0. Hence e_2 is the number of conjugacy classes of stabilizers G_α.

   Let A generate a G_α. Every modular matrix of order 2 is Γ-conjugate to $T = (0\,-1|1\,0)$. Hence $A = MTM^{-1}$, $M \in \Gamma$. Writing $\Gamma = \Gamma^0 \cdot \{B_i\}$, as in the text, we have $M = gB_i$ uniquely, for some $g \in \Gamma^0$ and some i in $1 \leqq i \leqq q+1$, and $A = gB_iTB_i^{-1}g^{-1}$. Clearly $B_iTB_i^{-1} \in \Gamma^0$. Hence every stabilizer G_α is associated uniquely with a matrix $B_iTB_i^{-1}$ in Γ^0. Moreover, if $G_\alpha = \{A\}$ is Γ^0-conjugate to $G_\beta = \{B\}$, then A and B are associated with the *same* $B_iTB_i^{-1}$. That is, there is a mapping from the conjugacy classes of G_α into the matrices $B_iTB_i^{-1}$ lying in Γ^0.

   Conversely, every $B_iTB_i^{-1} \in \Gamma^0$ is of order 2 and has the fixed point $\alpha = B_i(z)$, where z is the complex number i; hence it is the image under the mapping of the stabilizer G_α and so of the conjugacy class of G_α.

   The preceding remarks justify the assertion that *e_2 is the number of i, $1 \leqq i \leqq q+1$, for which $B_iTB_i^{-1}$ lips in Γ^0.*

   Let us use this result to calculate e_2. We have $B_j = S^j$, $1 \leqq j \leqq q$, $B_{q+1} = T$. Now

$$S^jTS^{-j} = \begin{pmatrix} \cdot & -j^2 & -1 \\ \cdot & & \cdot \end{pmatrix};$$

   hence this matrix is in Γ^0 if and only if $j^2 \equiv -1 \pmod q$, i.e., $1 + \left(\frac{-1}{q}\right)$ times. Since $TTT^{-1} = T$ is certainly not in Γ^0, we have

$$e_2 = 1 + \left(\frac{-1}{q}\right).$$

By a slightly more complicated discussion one can prove that

$$e_3 = 1 + \left(\frac{-3}{q}\right).$$

Substituting these results in equation (*) of the text, we find, for the genus $g = g(q)$ of $\Gamma^0(q)$:

$$g = \frac{q-6}{12} - \frac{1}{4}\left(\frac{-1}{q}\right) - \frac{1}{3}\left(\frac{-3}{q}\right), q = \text{prime}.$$

This formula shows at once that $q > 13$ implies $g > 0$ and that $g(11) = 1$.

2. We must show that $\Phi_q = \Phi$ is invariant on $\Gamma_0(q)$. Since Γ_0 is assumed to be of genus 0, there exists on Γ_0 a univalent function $\psi(\tau)$. We may assume ψ has a simple zero at i^∞ and a simple pole at 0, otherwise replace ψ by $(\psi(\tau) - \psi(i^\infty))/(\psi(\tau) - \psi(0))$ with obvious modifications if ψ has the zero but not the pole, etc. Furthermore, by multiplication by a constant we obtain

$$\psi(\tau) = x + \ldots, x = e^{2\pi i \tau}$$

as the expansion at i^∞. What we must now prove is that

$$\psi = \Phi.$$

Since ψ is univalent and has its zero and pole at i^∞ and 0, respectively, that is, at the cusps of the fundamental region R of Γ_0, it follows that ψ is regular and zero-free inside R.

Next, consider ψ^{q-1}; it is regular in R, invariant on Γ_0, has a pole of order $q-1$ at 0, and its expansion at i^∞ is $x^{q-1} + \ldots$. We assert the same, moreover, of the function

$$\varphi(\tau) = \Delta(q\tau)/\Delta(\tau),$$

where $\Delta = \eta^{24}$ is the function discussed in Chapter I, (26). As mentioned there, $\Delta \in \{\Gamma, -12\}$; hence $\Delta(q\tau) \in \{\Gamma_0(q), -12\}$. Therefore φ is invariant $\Gamma_0(q)$. Since $\Delta|T = \Delta$, we get

$$\varphi(\tau) = \frac{(\Delta|T)(q\tau)}{\Delta(\tau)|T} = \frac{(q\tau)^{-12}\Delta(-1/q\tau)}{\tau^{-12}\Delta(-1/\tau)} = q^{-12}x_1^{-(q-1)} + \ldots,$$

where $x_1 = e^{-2\pi i/q\tau}$ is a local variable at $\tau = 0$. Hence φ has a pole of order $q-1$ at $\tau = 0$. It is immediately clear from the definition that

$$\varphi(\tau) = x^{q-1} + \ldots$$

at $\tau = i^\infty$. Since Δ is regular and zero-free in R (in fact, in $\mathbf{H}$), φ is regular in $\mathbf{H}$. We have therefore established that φ and ψ^{q-1} have in common the properties mentioned above.

The function φ/ψ^{q-1} is now seen to be regular in R and at the cusps of R, namely, i^∞ and 0. Therefore φ/ψ^{q-1} is a constant (ch. I, sec. 2, th. 3), and this constant can only be 1 because the expansions of φ and ψ^{q-1} agree at i^∞. Since the two functions never vanish in $\mathbf{H}$, we can extract the $(q-1)$st root:

$$\psi(\tau) = \epsilon \cdot \{\eta(q\tau)/\eta(\tau)\}^r = \epsilon\Phi, \quad r = 24/(q-1)$$

where $\epsilon^{q-1} = 1$. Clearly ϵ is a regular function of τ in $\mathbf{H}$, but $|\epsilon| = 1$ implies $\epsilon =$ constant, and for the same reason as before, $\epsilon = 1$. Hence $\psi = \Phi$, as promised.

3. We may prove the formula for $\tau = iy$, $y > 0$ and then extend the result to all $\tau \in \mathbf{H}$ by analytic continuation. Taking logarithms we obtain ($\log y$ real)

$$\log \eta(i/y) - \log \eta(iy) = \tfrac{1}{2}\log y,$$

and substituting (3) we get

$$\sum_1^\infty \frac{1}{l}\frac{1}{1-e^{2\pi i y}} - \sum_1^\infty \frac{1}{l}\frac{1}{1-e^{2\pi l/y}} - \frac{\pi}{12}\left(y - \frac{1}{y}\right) = -\frac{1}{12}\log y \qquad (*)$$

as the formula to be proved.

Let

$$F_n(z) = -\frac{1}{4iz}\coth \pi N z \cot \frac{\pi N z}{y}, \; N = n + \frac{1}{2}, n = 1, 2, \ldots$$

and let C be the parallelogram joining the points y, i, $-y$, $-i$ in order. In C, F_n has simple poles at $z = il/N$ and at $z = -ly/N$, $l = \pm 1, \ldots, \pm n$. The residue at $z = il/N$ equals $(1/4\pi l)\cot \pi i l/y$, and noticing that this is even in l, we find

$$\sum_{l=-n}^{n}{}' \operatorname{Res} F_n(il/N) = \frac{1}{2\pi i}\sum_{l=1}^{n}\frac{1}{l} - \frac{1}{\pi i}\sum_{l=1}^{n}\frac{1}{l}\frac{1}{1-e^{2\pi l/y}}.$$

In the same way we get

$$\sum_{l=-n}^{n}{}' \operatorname{Res} F_n(-ly/N) = \frac{i}{2\pi}\sum_{l=1}^{n}\frac{1}{l} - \frac{i}{\pi}\sum_{l=1}^{n}\frac{1}{l}\frac{1}{1-e^{2\pi l y}}.$$

Besides these there is a triple pole at $z = 0$ with residue $i(y - y^{-1})/12$. Hence the sum of all residues of F_n in C is an expression whose limit for $n \to \infty$ equals $1/\pi i$ times the left member of (*). The proof will be complete, therefore, if we show that

$$\lim_{n\to\infty}\int_C F_n(z)dz = -\log y.$$

On the sides of C except at the vertices, $zF_n(z)$ has, as $n \to \infty$, the limit 1/4 on the sides connecting y, i and $-y$, $-i$, and the limit $-1/4$ on the other sides.

Furthermore, $F_n(z)$ is uniformly bounded on C for all n (use the fact that N is bounded away from the integers and that $y > 0$). Hence by Lebesgue's theorem of bounded convergence,

$$\begin{aligned}
\lim \int_C F_n(z)dz &= \frac{1}{4}\left\{-\int_{-i}^{y} + \int_{y}^{i} - \int_{i}^{-y} + \int_{-y}^{-i}\right\}\frac{dz}{z} \\
&= \frac{1}{2}\left\{-\int_{-i}^{y} + \int_{y}^{i}\right\}\frac{dz}{z} = \frac{1}{2}\left\{-\left(\log y + \frac{\pi i}{2}\right) + \left(\frac{\pi i}{2} - \log y\right)\right\} \\
&= -\log y,
\end{aligned}$$

as required.

Chapter VI

The Fourier Coefficients of $j(\tau)$. II.

1 Introduction

In the preceding lectures we have derived congruences for the Fourier coefficients $c(n)$ of the modular invariant $j(\tau)$ to the moduli 2, 3, 5, 7 and their powers. These primes, together with 13, are characterized by the property that they are the primes q for which $\Gamma_0(q)$ is of genus 0. But there is no congruence for 13. In particular it is a fact that $c(13) \not\equiv 0 \pmod{13}$.

On the other hand Morris Newman has proved (*Proc. Amer. Math. Soc.* **9** (1958), 609–612):

$$c(13np) + c(13n)c(13p) + p^{-1}c(13n/p) \equiv 0 \pmod{13}, \tag{1}$$

with $p =$ prime, $p \neq 13$, p^{-1} satisfies $pp^{-1} \equiv 1 \pmod{13}$, and as usual $c(x) = 0$ if x is not an integer. Now it so happens that $c(91) \equiv 0 \pmod{13}$. Hence choosing $p = 7$ we get

$$c(91n) \equiv 0 \pmod{13} \text{ if } (n, 7) = 1. \tag{2}$$

The last equation shows that $c(n)$ is divisible by 13 on a set of integers that contains arithmetic progressions. But (1) shows more than this. In fact, $c(13) \equiv -1 \pmod{13}$; and if we set

$$t(n) = -c(13n) \equiv c(13n)/c(13) \pmod{13},$$

then (1) becomes

$$t(np) - t(n)t(p) + p^{-1}t(n/p) \equiv 0 \pmod{13},\ p \neq 13. \tag{3}$$

The left member appears in the Hecke theory of forms of negative degree, which we discussed in Chapters II and III. Thus we can say that the Fourier coefficients of $j(\tau)$ exhibit the Hecke multiplicative relations (with $k = 0$) *as congruences to the modulus 13* rather than as identities.

Similar assertions can be made of the earlier moduli $q = 2, 3, 5, 7$ if we define $t(n) \equiv c(qn)/c(q) \pmod{q}$. This is possible, since it can be shown that $c(q)$ is divisible by q to the exact power $q^{r/2-1}$, $r = 24/(q-1)$, whereas $c(qn)$ is divisible by at least the same power; hence $t(n)$ is integral mod q. *Thus it appears that* $q = 13$ *is the first typical case*, the behaviour for the earlier values of q being obscured by the accident that $c(q) \equiv 0 \pmod{q^{r/2-1}}$.

In this chapter we shall prove the following theorems.

Theorem 1. *With* $p =$ *prime* $\neq 13$ *and*

$$t_\alpha(n) = t(n) \equiv c(13^\alpha n)/c(13^\alpha) \pmod{13^\alpha},$$

we have

$$t(np) - t(n)t(p) + p^{-1}t(n/p) \equiv 0 \pmod{13^\alpha}.$$

Newman also proved

$$c(13^2 n) \equiv 8c(13n) \pmod{13},$$

which can be generalized to

Theorem 2. *For all* $\alpha \geqq 1$ *there exists a constant* k_α *not divisible by* 13 *such that*

$$c(13^{\alpha+1}n) \equiv k_\alpha c(13^\alpha n) \pmod{13^\alpha},\ n = 1, 2, \ldots.$$

This may be regarded as a ramified form of theorem 1 for the prime $p = 13$.

The material of this chapter is based on two papers: A.O.L. Atkin and J. N. O'Brien, *Trans. Amer. Math. Soc.*, 1967, and a paper by Atkin as yet unpublished.

2 Basic Lemmas

Since most of the lemmas required here are analogous to those developed in the last chapter and are proved by the same techniques, we shall mostly confine ourselves to their statements.

The Hauptmodul on $\Gamma_0(13)$ is

$$\Phi_{13}(\tau) = \Phi = \left(\frac{\eta(13\tau)}{\eta(\tau)}\right)^2,$$

and we have

$$\Phi_{13}|W = 13^{-1}\Phi^{-1},\ W\tau = -1/13\tau. \tag{4}$$

Let

$$Y(\tau) = 13\Phi(\tau/13); \tag{5}$$

then

$$Y^{13} + \sum_{j=1}^{13} (-1)^j p_j Y^{13-j} = 0, \tag{6}$$

with

$$(-1)^{j+1} p_j = 13^3 \sum_{l=j}^{13} b_l \Phi^{l-j+1}, \tag{7}$$

j	1	2	3	4	5	6	7	8	9	10	11	12	13
$\pi(b_j)$	−1	0	1	2	4	4	5	6	7	8	9	10	10

(8)

Here π is the valuation defined by

$$\pi(a) = \pi_{13}(a) = e \text{ if } 13^e | a,\ 13^{e+1} \nmid a.$$

Note that

$$\pi(b_j) \geqq j - 2 + \left[\frac{-j+12}{14}\right] = \left[\frac{13j-16}{14}\right]. \tag{9}$$

Set

$$S_h = \sum_{l=0}^{13} Y_l^h = \sum_{k \geqq 1} c_{hk} \Phi^k, \tag{10}$$

where $Y_l = Y(\tau + l) = 13\Phi((\tau + l)/13)$, $l = 0, 1, \ldots, 12$ are the roots of (6). Then with $U_{13} = U$, we define a_{hk} by

$$\Phi^h | U = 13^{-1-h} S_h = \sum_{k \geqq 1} a_{hk} \Phi^k. \tag{11}$$

Lemma 1.

$$\pi(c_{hk}) \geqq \left[\frac{13(k+h+1)}{14}\right].$$

Proof. We have $S_1 = p_1 = 13^3 \sum_1^{13} b_k \Phi^k = \sum_{k \geqq 1} c_{1k} \Phi^k$ and so $c_{1k} = 13^3 b_k$, and by (9)

$$\pi(c_{1k}) \geqq 3 + \left[\frac{13k-16}{14}\right] = \left[\frac{13k+26}{14}\right],$$

as required.

Next assume the lemma for an $h-1$ and all $k \geqq 1$. Then using Newton's formula for the sum of powers of the roots of an algebraic equation (ch. V, (25)) we get

$$S_h = 13^3 \sum_{t\geqq 1} \Phi^t \sum_{j\geqq 1} \sum_{k\geqq 1} c_{h-j,k} b_{t-k+j-1} + 13^3 h \sum_{t\geqq 1} b_{t+h-1} \Phi^t,$$

which implies

$$c_{ht} = 13^3 h b_{t+h-1} + 13^3 \sum_{j\geqq 1} \sum_{k\geqq 1} c_{h-j,k} b_{t-k+j-1}. \tag{12}$$

The inequality

$$\left[\frac{a}{14}\right] + \left[\frac{b}{14}\right] \geqq \left[\frac{a+b-13}{14}\right] \tag{13}$$

is proved by noticing that it is periodic in a with period 14 for fixed b, and that for $0 \leqq a < 14$ the left member equals $[b/14]$ whereas the right member is $\leqq [b/14]$. Recalling that $\pi(a+b) \geqq \min(\pi(a), \pi(b))$, we now get from (12), (9), and the inductive assumption:

$$\pi(c_{ht}) \geqq \min\left(3 + \left[\frac{13(t+h-1)-16}{14}\right], 3 + \psi(h,t)\right),$$

where

$$\begin{aligned} \psi(h,t) &\geqq \min_{j,k} \left(\left[\frac{13(h+k-j+1)}{14}\right] + \left[\frac{13(t-k-1+j)-16}{14}\right]\right) \\ &\geqq \min_{j,k} \left[\frac{13(t+h)-16-13}{14}\right] = \left[\frac{13(t+h)-29}{14}\right], \end{aligned}$$

by (13). Hence

$$\pi(c_{ht}) \geqq \min\left(3 + \left[\frac{13(t+h)-29}{14}\right], 3 + \left[\frac{13(t+h)-29}{14}\right]\right) = \left[\frac{13(t+h+1)}{14}\right],$$

which completes the induction and proof. Q.E.D.

Corollary.

$$\pi(a_{hk}) \geqq \left[\frac{13k-h-1}{14}\right].$$

This follows immediately from (11).

3 Proof of Theorem 1

Theorem 1 is implied by the following assertion:

$$(j(\tau)|U^\alpha)|T_p \equiv l_\alpha \cdot (j(\tau)|U^\alpha) \pmod{13^\alpha},\ p \neq 13,\ U = U_{13}. \tag{14}$$

Indeed, let us recall that

$$j|U^\alpha = \sum_1^\infty c(13^\alpha n)x^n;$$

applying T_p (cf. ch. II, th. 2), we get (remember the degree $-k$ of $j(\tau)$ is 0),

$$c(13^\alpha np) + p^{-1}c(13^\alpha n/p) \equiv l_\alpha c(13^\alpha n) \pmod{13^\alpha}. \tag{15}$$

If we knew that

$$c(13^\alpha) \not\equiv 0 \pmod{13}, \tag{16}$$

then (15) would yield $l_\alpha \equiv c(13^\alpha p)/c(13^\alpha) \equiv t_\alpha(p)$, and this gives theorem 1 at once.

In order to come to grips with (14) we first observe that both members of this equation are polynomials in Φ with no constant term. For $j|U^\alpha$ this is simply equation (29) of Chapter V. For $j|U^\alpha|T_p$ we can prove the result by showing that $\Phi^l|T_p$ is a polynomial in Φ. Now T_p maps a function invariant on $\Gamma_0(q)$ into another such (ch. II, th. 1). In view of Chapter V, lemma 6, then, we have to show only that $\Phi^l|T_p$, $l \geq 1$, has expansions at $r = i^\infty$ and $\tau = 0$ of the form

$$\Phi^l|T_p = \sum_{m=8}^\infty a(m)x^m,\ s \geq 0,\ x = e^{2\pi i\tau} \tag{*}$$

$$\Phi^l|T_pW = \sum_{m=t}^\infty b(m)x^m, \tag{**}$$

respectively. (The integrality of $a(m)$, $b(m)$ is not needed since we are not insisting on the integrality of the C_l.) Now $\Phi^l = x^l + \ldots$, so (*) is evident. Since W commutes with T_p, we have, by (4),

$$\Phi^l|T_p|W = (\Phi^l|W)|T_p = 13^{-l}\Phi^{-l}|T_p = \left(\sum_{h\geq -l} d_h x^h\right)\Bigg|T_p = \sum_{m\geq -lp} b(m)x^m,^1$$

proving (**). Hence $j|U^\alpha|T_p$ is a polynomial in Φ with no constant term.

We now write

$$(j|U^\alpha)|T_p = \sum_{k\geq 1} g_{\alpha+1,k}\Phi^k,\ j|U^\alpha = \sum_{k\geq 1} j_{\alpha k}\Phi^k \tag{17}$$

which defines the numbers $g_{\alpha k}$, $j_{\alpha k}$. Then (14) is equivalent to

$$g_{\alpha+1,k} \equiv l_\alpha j_{\alpha k} \pmod{13^\alpha}. \tag{18}$$

Now set

$$\gamma_{st}^\alpha = g_{\alpha+1,s} j_{\alpha t} - g_{\alpha+1,t} j_{\alpha s},\ \alpha, s, t \geq 1. \tag{19}$$

If

$$\pi(\gamma_{st}^\alpha) \geq \alpha, \alpha \geq 1 \tag{20}$$

then (18) is true, provided (16) is proved. For we can choose $s = k$, $t = 1$; since $j_{\alpha 1} = c(13^\alpha) \not\equiv 0 \pmod{13}$ by (16), the relation (20) yields (18) with $l_\alpha = j_{\alpha 1}^{-1} g_{\alpha+1,1}$.

Hence *what we have to prove are* (20) *and* (16).

Proof of (20). The proof of (20) depends on the following result:

Lemma 2. *For* $\alpha \geq 1$,

$$\pi(j_{\alpha k}) \geq \left[\frac{13k-2}{14}\right], \pi(g_{\alpha+1,k}) \geq \left[\frac{13k-2}{14}\right].$$

Lemma 2, in turn, is proved by induction on α. The case $\alpha = 1$ is

Lemma 3.

$$\pi(j_{1k}) \geq \left[\frac{13k-2}{14}\right], \pi(g_{2k}) \geq \left[\frac{13k-2}{14}\right].$$

The first assertion follows from Chapter V, (15), since $j_{1k} = C_k 13^{k-1}$. To prove the second, we need two other lemmas.

Lemma 4. *Setting*

$$\Phi|T_p = \sum_{k\geq 1} f_k \Phi^k$$

we have

$$\pi(f_k) \geqq k - 1.$$

Proof of Lemma 4. Let $W\tau = -1/13\tau$; W commutes with all T_p, $p \neq 13$. Hence using (4) and Note 1,

$$\begin{aligned}\Phi|T_p|W = \Phi|W|T_p &= 13^{-1}\Phi^{-1}|T_p = 13^{-1}\{x^{-1} + b_0 + \ldots\}|T_p \\ &= 13^{-1}p^{-1}\{x^{-p} + b_0' + \ldots\} \\ &= 13^{-1}p^{-1}\{\Phi^{-p} + e_{p-1}\Phi^{-p+1} + \ldots + e_1\Phi^{-1} + b_0''\},\end{aligned}$$

with $e_j \in Z$ since Φ^{-1} is over Z. Replacing τ by $-1/13\tau$ (i.e., multiplying by W^{-1}) replaces ϕ^{-1} by 13Φ, so

$$\Phi|T_p = 13^{-1}p^{-1}\{13e_1\Phi + 13^2e_2\Phi^2 + \ldots + 13^p\Phi^p\}. \qquad \text{Q.E.D.}$$

Lemma 5.

$$j|U = -\Phi + 13^2\Phi|U^2.$$

We sketch the proof. The function

$$F(\tau) = 13\{\Phi(-1/13\tau) + 13\Phi|U\} = \Phi^{-1} + 13^2\Phi|U$$

– the identity of the two right members is clear – can be shown to be in $\{\Gamma, 0\}$. Hence the function

$$G(\tau) = j(\tau) - (\Phi^{-1}(\tau) + 13^2\Phi(\tau))|U$$

is in $\{\Gamma, 0\}$, and besides is regular at i^∞. Therefore $G =$ constant. Apply U:

$$j|U = \Phi^{-1}|U + 13^2\Phi|U^2 + \text{const.}$$

Next consider

$$H(\tau) = \Phi^{-1}(-1/13\tau) + 13\Phi^{-1}(\tau)|U = 13\Phi(\tau) + 13\Phi^{-1}(\tau)|U;$$

as before $H \in \{\Gamma, 0\}$, also H is regular at i^∞ and so constant. Thus

$$\Phi^{-1}|U = -\Phi + C. \qquad \text{Q.E.D.}$$

We shall now complete the proof of lemma 3. Apply T_p to both sides in lemma 5; since T_p commutes with U (cf. ch. III, sec. 2, lemma 9), we get

$$j|UT_p = -\Phi|T_p + 13^2\Phi|T_p|U^2 = -\sum_{k\geqq 1} f_k\Phi^k + \sum_{k\geqq 1} h_k\Phi^k. \tag{21}$$

First, recalling (11),

$$\Phi|T_p|U = \sum_{\rho\geq 1} f_\rho\Phi^\rho|U = \sum_{\rho\geq 1} f_\rho \sum_{k\geq 1} a_{\rho k}\Phi^k = \sum_{k\geq 1}\Phi^k \sum_{\rho\geq 1} f_\rho a_{\rho k}.$$

By lemma 4 and the corollary to lemma 1, if we set

$$\Phi|T_pU = \sum_{k\geqq 1} m_k\Phi^k,$$

then

$$\begin{aligned}\pi(m_k) = \pi\left(\sum_{\rho\geq 1} f_\rho a_{\rho k}\right) &\geqq \min_\rho\left\{(\rho-1) + \left[\frac{13k-\rho-1}{14}\right]\right\}\\ &= \left[\frac{13k-2}{14}\right],\end{aligned}$$

since the expression under the minimum is nondecreasing as $\rho \to \rho+1$. Next

$$\Phi|T_p|U^2 = \sum_{\rho \geqq 1} m_\rho \Phi^\rho|U = \sum_{k \geqq 1} n_k \Phi^k,$$

and

$$\pi(n_k) \geqq \pi\left(\sum_{\rho \geqq 1} m_\rho a_{\rho k}\right)$$
$$\geqq \min_\rho\left(\left[\frac{13\rho-2}{14}\right]+\left[\frac{13k-\rho-1}{14}\right]\right) = \left[\frac{13k-2}{14}\right].$$

Hence from (21),

$$\pi(h_k) = 2+\pi(n_k) \geqq \left[\frac{13k+26}{14}\right].$$

Recalling (17) and lemma 4 we finally get that

$$\pi(g_{2k}) \geqq \min(\pi(f_k), \pi(h_k)) \geqq \min\left(k-1, \left[\frac{13k+26}{14}\right]\right).$$

Since $k-1 \geqq k+\left[\frac{-k-2}{14}\right] = \left[\frac{13k-2}{14}\right]$, it is clear that whatever the minimum may be, $\pi(g_{2k}) \geqq \left[\frac{13k-2}{14}\right]$. This completes the proof of lemma 3.

Next we establish the inductive step $\alpha \to \alpha+1$ in lemma 2. Assume the result for some α and all $k \geqq 1$. The proof is the same for the two functions, so let $h_{\alpha k}$ be either $g_{\alpha+1,k}$ or $j_{\alpha k}$. The inductive assumption is then $\pi(h_{\alpha k}) \geq \left[\frac{13k-2}{14}\right]$. We have, by (17), (11), and the commutativity of T_p and U,

$$\sum_{k \geqq 1} h_{\alpha+1,k}\Phi^k = \sum_k h_{\alpha k}\Phi^k|U = \sum_k h_{\alpha k}\sum_l a_{kl}\Phi^l,$$

so that using the corollary to lemma 1 we get

$$\pi(h_{\alpha+1,k}) = \pi\left(\sum_{l \geqq 1} h_{\alpha l}a_{lk}\right) \geqq \min_{l \geqq 1}(\pi(h_{\alpha l})+\pi(a_{\alpha l})) \tag{22}$$
$$\geqq \min_{l \geqq 1}\left(\left[\frac{13l-2}{14}\right]+\left[\frac{13k-l-1}{14}\right]\right) \geqq (\quad)_{l=1} = 0+\left[\frac{13k-2}{14}\right].$$

by the usual argument involving $l \to l+2$. We have proved lemma 2.

We are going to use lemma 2 to prove (20), but we can do this only by proving a stronger result.

Lemma 6.

$$\pi(\gamma_{sl}^\alpha) \geqq \alpha+\left[\frac{13(s+t)-32}{14}\right], \alpha \geqq 1; s,t \geq 1, s+t \geq 3.$$

The last restriction on s, t is valid since $\pi(\gamma_{11}^{\alpha}) = 0$. For $\alpha = 1$:

$$\gamma_{sl}^{1} = g_{2s} j_{1t} - g_{2t} j_{1s},$$
$$\pi(\gamma_{sl}^{1}) \geqq \min\{\pi(g_{2s}) + \pi(j_{1t}), \pi(j_{1s}) + \pi(g_{2t})\},$$

and by lemma 3 both terms are not less than

$$\left[\frac{13s-2}{14}\right] + \left[\frac{13t-2}{14}\right] \geqq \left[\frac{13(s+t)-4-13}{14}\right] \geqq 1 + \left[\frac{13(s+t)-31}{14}\right],$$

where we have used the inequality (13).

For $\alpha > 1$ we have, as in (22),

$$h_{\alpha+1,k} = \sum_{l \geqq 1} h_{\alpha l} a_{lk}, \; h_{\alpha k} = g_{\alpha+1,k} \text{ or } j_{\alpha k};$$

so that

$$\begin{aligned} \gamma_{st}^{\alpha+1} &= g_{\alpha+2,s} j_{\alpha+1,t} - j_{\alpha+1,s} g_{\alpha+2,t} \\ &= \sum_{l \geqq 1} g_{\alpha+1,l} a_{ls} \sum_{m \geqq 1} j_{\alpha m} a_{ml} - \sum_{l \geqq 1} j_{\alpha+l} a_{ls} \sum_{m \geqq 1} g_{\alpha+1,m} a_{ml} \\ &= \sum_{l,m} a_{ls} a_{mt} \{g_{\alpha+1,l} j_{\alpha m} - j_{\alpha l} g_{\alpha+1,m}\} = \sum_{l,m} \gamma_{lm}^{\alpha} a_{ls} a_{mt}. \end{aligned}$$

In this equation note that $l + m \geqq 3$, for $l + m = 2$ only for $l = m = 1$ and $\gamma_{11}^{\alpha} = 0$. Hence, applying the inductive assumption, we get, using (13) and the inequalities on $\pi(a_{hk})$,

$$\pi(\gamma_{st}^{\alpha+1}) \geqq \min_{\substack{l,m \\ l+m \geqq 3}} \left\{\alpha + \left[\frac{13(l+m)-32}{14}\right] + \left[\frac{13(s+t)-(l+m)-15}{14}\right]\right\}.$$

The right member is increasing for $l + m \to l + m + 2$, and $l + m \geqq 3$; its minimum is attained for $l + m = 3$. Therefore

$$\pi(\gamma_{st}^{\alpha+1}) \geqq \alpha + \left[\frac{13(s+t)-18}{14}\right] = \alpha + 1 + \left[\frac{13(s+t)-32}{14}\right].$$

This concludes the proof of (20).

Proof of (16). We wish to prove

$$\pi(j_{\alpha 1}) = 0, \quad \alpha \geqq 1. \tag{16}$$

Let us assume this for an α, then

$$j_{\alpha+1,1} = \sum_{\rho \geqq 1} j_{\alpha\rho} a_{\rho 1} = j_{\alpha 1} a_{11} + \sum_{\rho \geqq 2} j_{\alpha\rho} a_{\rho 1}, \tag{23}$$

where a_{hk} is, as before, defined by

$$\Phi^h | U = \sum_{k \geqq 1} a_{hk} \Phi^k.$$

Recall that $\pi(a+b) = \min(\pi(a), \pi(b))$ whenever $\pi(a) \neq \pi(b)$, Now $\pi(a_{11}) = 0$. Indeed, from (7) and (8), $S_1 = p_1 = 13^3(b_1\Phi + \ldots) = 13^2(13b_1\Phi + \ldots)$ and $\pi(13b_1) = 0$. Since $S_1 = \sum_{k \geqq 1} c_{1k}\Phi^k$, this gives $\pi(c_{11}) = 2$, and $a_{11} = 13^{-2}c_{11}$, which implies $\pi(a_{11}) = 0$. Hence $\pi(j_{\alpha 1} a_{11}) = \pi(j_{\alpha 1}) + \pi(a_{11}) = 0 + 0 = 0$, by the inductive assumption. So it follows from (23) that it is sufficient to prove

$$\pi\left(\sum_{\rho > 1} j_{\alpha\rho} a_{\rho 1}\right) > 0. \tag{24}$$

But

$$\pi\left(\sum_{\rho > 1} j_{\alpha\rho} a_{\rho 1}\right) \geqq \min_{\rho > 1}\{\pi(j_{\alpha\rho}) + \pi(a_{\rho 1})\}$$

$$\geqq \min_{\rho > 1}\left\{\left[\frac{13\rho - 2}{14}\right] + \left[\frac{13 - \rho - 1}{14}\right]\right\},$$

and remembering $\rho \geqq 2$, this minimum is attained for $\rho = 2$ and the minimum $= 1 > 0$. Hence (24) is proved and therefore (16) for $\alpha + 1$.

Finally, we must verify for $\alpha = 1$. But this is clear from lemma 5; in fact,

$$j_{11} \equiv -1 \pmod{13^2}.$$

This completes the proof of theorem 1. Theorem 2 is proved by the same methods but the proof is easier. Q.E.D.

Chapter VI Notes

1. In Chapter II, section 2, theorem 2 we developed a formula for the effect of T_p on a Fourier series starting with a term in x^2 *with* $s \geq 0$. The extension to $s < 0$ is easily made and yields:

$$\sum_{m \geq s} a(m)x^m | T_p = \sum_{mp \geq s} a(mp)x^m + p^{-1} \sum_{m \geq ps} a(m/p)x^m = \sum_{m \geq ps} a'(m)x^m.$$